DUXFORD

Airfield of Dreams

DUXFORD
Airfield of Dreams

A GUIDE TO AIRCRAFT
AT DUXFORD

John M. Dibbs

Text by David Legg

Airlife
England

For Olga and Ken.

Acknowledgements:
The list of people to whom I owe thanks is long, and I have probably missed out lots of people, but here are a few that I can honestly say I could not have done this book without:
OFMC — Mark, Ray and Sarah Hanna, Roger Shepherd, Alan Walker, Brian Smith, Rolf Meum, Rod Dean.
TFC — Stephen Grey, Nick Grey, Hoof Proudfoot, Pete Rushen, Jeanne Frazer, Pippa Vaughan, Paul Bonhomme.
Radial Pair/Yak — Norman Lees, Gary Numan, Eddie Coventry and BAC Windows.
Plane Sailing Air Displays/Fighter Meet — Paul Warren Wilson, Arthur Gibson, Mike Carter.
TARC — Graham Warner, John Romain, Lee Proudfoot.
Trevor Butcher, Carolyn Grace, Pete Kynsey, Richard Parker.
B-17 Preservation Ltd — Eli Sallingboe, Keith Sissons.
The Imperial War Museum staff — David Henchie, Andrea Hamblin, Frank Crosby, Carol Stearn, Mike Zweiff.
109 Restoration Team — Russ Snadden, John Elcombe, John Allison, Dave Southwood.
Historic Flying, Audley End — Clive Denney, Tim Routsis.
Warbirds Worldwide — Paul Coggan.
Lindsey Walton Collection — Lindsey Walton.

Special thanks to my mother and father for putting up with me for the two-and-a-half years it took to photograph this volume, and for their unending encouragement and support. I also owe John Harbourne a pint.

I would like to thank especially John Thirkettle, Brian Jones, Steve Baker and staff at Leeds Cameras (Photovisual) London for their tireless patience and technical assistance.

This entire book was shot on Fujichrome 100 and 50 ASA stock, using Canon 35mm equipment.

Design: John Harbourne/Tracey Adam/J. Dibbs.
Captions: G. Numan/J. Dibbs.

First published in the UK in 1992 by
Airlife Publishing Ltd.

British Library Cataloguing in Publication Data
A catalogue record for this book
is available from the British Library.

ISBN 1 85310 227 X

Printed by Livesey Ltd., Shrewsbury.

Airlife Publishing Ltd.
101 Longden Road, Shrewsbury SY3 9EB, England

FOREWORD

By Gary Numan

Duxford. The name itself is legendary. An airfield as important to the history of Britain as the aircraft and men that once scrambled heroically from its grass, hurling themselves with unswerving courage against the awesome might of the German Luftwaffe.

Duxford. Now firmly and justifiably established as one of the most important warbird centres in the world today. It helped to make history, and it now helps to keep that history alive. This book aims to celebrate that remarkable achievement and I can think of no better man to be given the task than John Dibbs.

I have come to know John very well during the making of this book. I flew many trips as pilot of one of the camera ships and many more as pilot of subject aeroplanes, both my own Harvard and the BAC Window's immaculate Yak 11. I also had a small hand in helping to write the captions. This rather unique involvement gave me an insight not only into the making of a work such as this, but also into the enormous care and sheer hard effort that John has gone to, to make this the finest book of its kind that I have ever seen. When you learn that this is John's first ever book to be published, the achievement seems all the more remarkable. I genuinely believe him to be one of the most talented and innovative aviation photographers in the world today.

I clearly remember, on one occasion, seeing John hanging, unstrapped, from the back of the camera ship, being battered mercilessly by the bitterly cold slipstream and yet still giving clear and positive instructions to me in the Yak, as well as hanging on to two cameras and saving a set of goggles torn off in the 200 mph blast. I was to learn that this kind of thing was quite a normal Dibbs tactic. His pre-shoot planning is quite meticulous and his enthusiasm for aviation is total. No wonder that so few other photographers manage to achieve such amazing results.

To work with, he is good fun, always late, extremely conscientious and sincerely grateful to the pilots and owners who often give up time and money to help on a project like this. I am pleased to say that we became good friends during the making of *Duxford. Field of Dreams*. It was always a great pleasure for me and I enjoyed every minute of it.

AUTHOR'S NOTE

This book is intended to be a guide to the aircraft based at Duxford, both static and airworthy. Such is the wealth of the Duxford heritage that it is not possible to show pictorially all the types. Rather, this is a collection of photographs illustrating the most exciting aspects of historic aircraft preservation and operation — backed by a complete and informative text.

A BRIEF HISTORY

As the world aircraft preservation movement has developed over the years, many museums of varying sizes and quality have become established, some in large, enclosed premises, others in outdoor locations. The largest of the museums have become internationally famous — The Royal Air Force Museum at Hendon, the National Air and Space Museum in Washington D.C., the United States Air Force Museum at Dayton, Ohio and the Musee de L'Air in Paris spring to mind. Other locations have become well known as centres of vintage aircraft rebuilding. In the USA, Chino in California has become something of a vintage aircraft enthusiast's Mecca, whilst La Ferté Alais in France is well known for its exotic collection of aircraft and cache of airframes awaiting restoration. Finally, certain airfields build a reputation for airshows, such as North Weald in Essex and its Fighter Meet, West Malling in Kent for The Great Warbirds Air Display, Harlingen in Texas for the Confederate Air Force shows prior to the CAF's move of base, and Oshkosh, Wisconsin and its mammoth annual gathering of vintage aircraft and home-builts. Old Warden in Cambridgeshire is synonymous with vintage airshows as well as its collection of vintage aeroplanes.

There are few places than can claim to combine a large indoor and outdoor collection of aircraft, substantial numbers of restoration projects of all shapes and sizes and a regular display venue in one facility, but the Imperial War Museum site at Duxford in Cambridgeshire does just that. Indeed, Duxford has justifiably become a focus for much that happens in the world vintage aircraft scene.

It is entirely appropriate therefore that this airfield, which houses such a magnificent collection of historic aircraft, military equipment and artifacts, should in itself be an historic site with a past that is almost as old as military aviation and slightly older than the Royal Air Force.

The construction of a military air station at Duxford commenced whilst the Royal Flying Corps and Royal Naval Air Service were still in existence, although it did not open until after these two services had been amalgamated to form the Royal Air Force on April 1, 1918. However, it had its origins in the First World War and came into being as a result of expansion of the Royal Flying Corps and consequent requirement for more trained airmen. Ironically, particularly in view of events many years later, it was trainee American servicemen that first used Duxford, the initial personnel arriving in early 1918. Americans were to be stationed at Duxford again in the Second World War in some numbers and, in more recent years, many vintage American aircraft types have called Duxford home.

Following this, squadrons of the fledgeling Royal Air Force used Duxford as a home base after returning from war on the continent and prior to being disbanded at the cessation of hostilities.

Even at this early stage, the basic layout of the airfield that remains to this day was established with facilities straddling what is now the A505 road. The aerodrome proper, with its dispersal areas and hangarage, was situated to the south, and three of the magnificent wooden hangars with their latticework 'Belfast Trusses' in the roof and concertina-type folding doors dating from this period, still survive.

By the time of its completion, the war for which Duxford's prospective trained airmen were destined was over and the peacetime Royal Air Force was facing a contraction in size which would result in squadrons and manpower being reduced to a fraction of their former numbers. This inevitably placed the future use of some RAF airfields in doubt.

Duxford owed its survival to the proximity of Cambridge and its University colleges as it was ideally placed to serve as a base for flight research by aeronautical engineering students. Additionally, in the spring of 1920, Duxford became home for 2 Flying Training School with its Avro 504s, Bristol F2B fighters and DH9As. Even a few surviving RE8s flew from there for a while, and later some Sopwith Snipes were added. Eventually, 2FTS moved on to Digby in Lincolnshire. Prior to this, the streamlined Royal Air Force began to expand slightly and RAF Duxford entered into its phase as a fighter aircraft station, a role it was to retain for nearly forty years.

The first units to be reformed there were 19, 29 and 111 Squadrons. 19 Squadron initially operated Sopwith Snipes and then, in the period up to 1937, was equipped with a succession of fighters that typified the wonderful inter-war biplane years — the Gloster Grebe, Armstrong Whitworth Siskin, Bristol Bulldog and the Gloster Gauntlet. 29 Squadron also flew Snipes, Grebes and Siskins until moving on to North Weald in April 1928, whilst 111 Squadron flew these same three types and then migrated to Hornchurch at the same time as 29 Squadron moved out.

Gloster Grebes *of No. 19 Squadron in 1927.*

One of Duxford's great pre-war moments took place on July 6, 1935 when it was chosen as the location for the review of the Royal Air Force by King George V in his Silver Jubilee year. The Royal dais was erected facing the runway to the south and the aircraft of twenty squadrons flew past in stately procession, watched by over 100,000 people.

Although Duxford had become fully established as a fighter station, its size allowed for other units to be stationed there. It was an obvious choice of base for the Cambridge University Air Squadron, which moved there in early 1936 and operated Avro 504s and, later, Avro Tutors. For a short period in 1936, Bristol Bulldogs and Gloster Gauntlets of the Meteorological Flight flew from Duxford also.

The sole fighter squadron remaining by the mid-1930s — 19 Squadron — was joined in July 1936 by 66 Squadron, and both operated Gloster Gauntlets until late 1938. However, the summer of 1938 saw the first example of that soon to be famous single-engined fighter, the Supermarine Spitfire, arrive at Duxford. In fact, the first Duxford Spitfire, serial K9792, was delivered on Saturday July 30, 1938 and it gave a display to the few assembled onlookers in the hands of Supermarine test pilot Jeffrey Quill. This aircraft was entrusted to 19 Squadron, which was to thus go down in history as the very first operational Spitfire Squadron, a distinction indeed. However, 66 Squadron was not far behind, with their first Spitfires arriving in the following October. By the time that Britain declared war on Germany, 611 Squadron was also resident at Duxford with its own complement of Spitfires, although they deployed to Digby by the end of October 1939.

A new shape in the skies around Duxford came in that month, when Bristol Blenheims began to arrive. The Mk IF equipped 222 Squadron in the fighter role. From early 1940, this squadron was commanded by Douglas Bader, then a Flight Lieutenant. Bader was no stranger to Duxford, having served there as a ground-based Flying Officer following his flying accident in which he lost both legs and, again, in 1939 as a 19 Squadron pilot.

By March 1940, 222 Squadron had exchanged its Blenheims for Spitfires, thus adding yet another squadron to the number that operated this most legendary fighter aircraft from Duxford. Throughout 1940, the

Spitfire *Press Day, 4 May 1939.*

premier Spitfire squadron, No. 19, moved to and fro between Duxford and its nearby 'satellite' airfield at Fowlmere.

Another new type moved to Duxford in May 1940, when 264 Squadron arrived with the Boulton Paul Defiant single-engined, turreted fighter. Their stay was short, however, and within three months they had moved on to Fowlmere, never to return.

When the Battle of Britain commenced in mid-1940, Duxford became a scene of great activity although, geographically, it was some way to the north of the main scene of action. Nonetheless, its aircraft were frequently scrambled to fight against the Luftwaffe.

Although a Spitfire base right from the start of the type's entry into service, Duxford did not see Hurricanes resident there until July 1940, when 310 (Czech) Squadron moved in. They were to stay until June the following year, when they transferred to Martlesham Heath. Prior to their move, however, 310's Hurricanes were joined by those of 242 Squadron in late October 1940, under the command of the returning Douglas Bader. Their stay was to be very brief, and by the end of November they had gone to Coltishall in Norfolk. Yet another squadron briefly based at Duxford and also operating Hurricanes was 312 (Czech) Squadron. Arriving in the late summer of 1940, they had moved to Speke (Liverpool) by the end of the following month. Such were the requirements of wartime operations!

Three of the Duxford squadrons, namely 19, 242 and 310, became a Wing under the command of Douglas Bader and pioneered the sometimes controversial 'Big Wing' approach to fighting large formations of incoming enemy aircraft. The strategy did meet with some success and as a result two further squadrons were added to the Wing, now known as 12 Group, these being 302 (Polish) Squadron and 611 (Auxiliary) Squadron, flying Hurricanes and Spitfires respectively. 611 had first formed at Duxford in August 1939. Not all of the aforementioned squadrons were permanently based at Duxford during the Big Wing phase, but used the airfield as an assembly point for their daily forays.

At the end of 1940, a new dimension was added to the activity at Duxford with the arrival of the Air Fighting Development Unit. This organisation was tasked with the evaluation of tactical developments for which purpose it was equipped with the obligatory Spitfires and Hurricanes. However, another major role of the AFDU was to test new aircraft as they became available. Up until its departure to Wittering in early 1943, the AFDU tested many diverse types, including such American equipment as the Martin Maryland, North American Mitchell, Martin Marauder and Lockheed Ventura, as well as British types such as the Handley Page Halifax bomber. Far from the sea, the Royal Navy also operated an evaluation unit at Duxford, testing mainly British types and known as the Royal Navy AFDU.

In the late summer of 1941, the AFDU acquired various captured Luftwaffe aircraft, including a Heinkel He III bomber, Messerschmitt Bf 109 and Bf 110. These were used for airborne evaluation and comparison against similar Allied types. Later, in November of the same year, 1426 (Enemy Aircraft) Flight was formed with the task of exhibiting captured

German aircraft to other flying units around the British Isles. Often unofficially known as the 'Rafwaffe', 1426 Flight eventually transferred to Collyweston in March 1943, having up to that time carried out a total of nine tours to airfields all around the country.

The Czech Hurricane squadron, No. 310, flew to its new base at Martlesham Heath in June 1941, as mentioned earlier, and was immediately replaced at Duxford by 56 Squadron, another Hurricane unit. Within a short space of time, 56 began to equip with early Hawker Typhoons, to be followed by 266 Squadron in January the following year and 609 Squadron two months after that, thus forming a Duxford Typhoon Wing. The Typhoon's progress toward an effective operational aircraft was a somewhat long and tortuous one but it eventually became a fine fighting machine, used on offensive sweeps over the Continent.

TFC's Hurricane *during rebuild.*

Meanwhile, testing of aircraft had still been going on with 601 Squadron having arrived in mid August 1941, with the unusual Bell Airacobra and its mid-fuselage mounted powerplant and cannon firing through the propeller spinner. The unit stayed only until the beginning of 1942, when it went to Acaster Malbis near Leeds. The Airacobra proved unsuitable for RAF use and lasted only a short while after the move from Duxford. It was replaced in 601 Squadron service by the Spitfire V. Although not a success as far as the RAF was concerned, some other Airacobras were based at Duxford for a while in 1942 in the hands of the 350th Fighter Group prior to deploying to North Africa. Lockheed P-38 Lightnings were also tested on trials at Duxford for a brief period in 1942.

A lesser known but nonetheless important unit at Duxford was 74 Signals Wing, which was based there in the early 1940s using Bristol Blenheims and Cierva Autogiros for radar calibration work.

In April of 1943, Duxford was handed over to the United States Eighth Air Force, all British operational units having moved out by then. Known as Base 357, it was designated as headquarters of the 78th Fighter Group and at first, large numbers of Republic P-47 Thunderbolts were resident. These large single-engined fighters were used to escort the massive United States Army Air Force bombing raids to Germany and France for low-level attacks against strategic targets and targets of opportunity. They were also heavily involved in providing air cover on D-Day. The 78th flew 'razor

back' and tear drop hooded versions of the 'Jug'. New equipment arrived in late 1944, when the Thunderbolts were replaced by P-51 Mustangs. One particular distinction was claimed by the 78th, that of being the first 8th Air Force unit to shoot down a Messerschmitt Me 262 jet fighter.

By the war's end, the 78th could claim a total of nearly 1,400 enemy aircraft destroyed either in the air or on the ground.

Up until the early winter of 1944, Duxford remained a grass airfield but at this time, the Americans laid a metal runway using the famous American Universal Pierced Steel Planking or PSP.

The 8th Air Force units finally moved out of Duxford for good in August 1945 and the airfield was handed over to the Royal Air Force on December 1, having been left under what is known as 'Care and Maintenance' status in the intervening period.

Perhaps fittingly considering the airfield's past history, the first RAF aircraft to return to Duxford were Spitfires! The Mk IX version equipped 165 Squadron when it arrived in January 1946, and these were to be followed by Mk XXIs in April when 91 Squadron moved in. Both squadrons were destined to leave again by the end of the year, but to return in 1947 with jets.

The first jet-equipped squadrons to be based at Duxford were in fact 92 and 56 Squadrons, with Meteor IIIs. Initially, the new generation aircraft flew from the old PSP runway, but by the summer of 1951 a new concrete runway and dispersals had been laid. From that time, Meteor F8s of 64 and 65 Squadrons became resident, whilst in September 1956 night fighter versions of the Meteor was based at Duxford with 64 Squadron. In order to house the increased number of jets, a metal T2 hangar was erected to the east of the four groups of original Belfast hangars. The latter were in fact to outlive the T2, although in more recent years a replacement has been built on the same site and is now used to house the aircraft of The Fighter Collection.

At the end of 1952, that beautiful aircraft, the Hawker Hunter, arrived for service with 65 Squadron and they remained until the squadron disbanded in March 1961. The Meteor NF 12s and NF 14s of 64 Squadron were replaced by the Gloster Javelin delta-winged fighter in September 1958, and both FAW 7 and FAW 8 variants were used until the squadron relocated to nearby Waterbeach in July 1961, the last operational squadron to be based at Duxford.

By now, Duxford's position was such that it was in the wrong place for the RAF's needs. The perceived threat was from the east, and the RAF front line fighter stations needed to be situated within a short distance of the east coast of England and Scotland. Duxford's future was not helped by the age of its facilities and the inability of the runway to accept the new generation of fighter — the English Electric, later BAC, Lightning.

The last operational RAF flight from Duxford was flown by a Meteor T7 on 31 July 1961, the pilot in command being Air Vice Marshal R. N. Bateson. Thereafter, the base remained unused, the dispersals silent and the hangars empty of aircraft. Its future seemed uncertain to say the least, and was to remain so for some fifteen years. During that time, however,

Duxford did have a brief return to its former glory days when it was used for some of the aerial activity connected with the motion picture *Battle of Britain*. This famous film involved air-to-air filming sequences featuring Spitfires and Hurricanes, as well as ex-Spanish Air Force aircraft. The latter were Spanish licence-built Messerschmitt 109s, Heinkel He 111s and Junkers 52s. The gathering together of a large number of these vintage wartime aircraft and, in many cases, their rebuild to flying condition, proved that it was still possible — just — to locate flyable Second World War aircraft. The effort to find these old machines by former RAF bomber pilot Hamish Mahaddie in some ways laid the foundation for the present day 'warbird' preservation movement in the United Kingdom and Europe.

Duxford became the main operational location for the aerial film work as it was ideal for the purpose with its 'period' hangars and buildings and unspoilt skyline to the south. The aircraft moved in during the spring of 1968. The south-west corner of the airfield even became, thanks to the film set technicians, a replica of a typical French aerodrome circa May 1940, complete with French chateau! During the filming, Duxford's large, empty hangers had proved particularly useful, but Hangar 3 was to become even more so, albeit in somewhat unorthodox fashion. In mid-June it was destroyed in a spectacular set-piece re-enactment of a bombing raid. Its demise can be seen in the film, where it is blown to smithereens amidst much flame and smoke. It is still a matter of debate as to whether permission for its destruction had been given in advance! The old hangar's concrete floor is still used as a hardstanding for static aircraft on display to this day. By late 1968, Duxford was once again deserted and silent.

Shortly after this brief return to life, the RAF decided to dispose of Duxford airfield and this was made official in 1969. Initial proposals to use the site as a prison, industrial storage area or regional sports centre came to naught. At this time, the Imperial War Museum was looking for a site to store and ultimately display aircraft in its collection which could not be kept at its London museum site in South London because of space considerations. With the newly-formed East Anglian Aviation Society, one hangar was prepared to receive aircraft and other exhibits. The first aircraft to arrive was a P-51D Mustang, whilst the first to arrive by air was an ex-Fleet Air Arm Sea Vixen, which flew in during March 1972. The first public air display was held in October of the following year and increasingly large and popular shows have been a regular feature ever since.

By the mid-1970s, Duxford had become a major part of the Imperial War Museum and in June 1976 it was opened to the general public on a daily basis. The number of visitors and exhibits has grown apace since then. In the late summer of 1977, the runway had to be shortened during the building of the adjacent M11 motorway, although this has not prevented large piston-engined types and some jets from operating there. In the same year, the Cambridgeshire County Council purchased the remaining runway and some of the airfield, thus securing its future.

In 1980, a 'new' T2 hangar was erected on the site of the original in order to give more covered space for aircraft, and some time afterwards the massive 'Superhangar' was erected in the north-east corner. This structure is capable of holding the largest of the aircraft based at Duxford and has enabled these larger exhibits to be rotated between indoor and outdoor display. It also acts as a spray-shop when these aircraft are repainted as part of their restoration. Over the years, a continual programme of renovation of the site and its buildings has progressed with the establishment of many permanent exhibits and imaginative 'diorama' type scenes in some hangars.

The administration and operation of Duxford is shared by several organisations. In addition to the Imperial War Museum and Cambridgeshire County Council, the Duxford Aviation Society (DAS) plays an important role in many practical ways through the efforts of its volunteer membership. As well as carrying out many tasks throughout the year, the DAS has also acquired a splendid collection of post-war British commercial transport aircraft, both propeller and jet powered.

Another feature of the Duxford scene is the number of private aircraft collections which are now based there and which add so much interest to an already fascinating place. Sadly, the first collection to move there is no more. Ormond Haydon-Baillie chose Duxford as the base for his expanding historic aircraft collection in the early 1970s and, within a short space of time, it consisted of three airworthy Lockheed T-33s, a Hawker Sea Fury, an Avro Canada CF-100 Canuck jet and the complete but dismantled parts of two Bristol Blenheims under long-term restoration. In 1977, Ormond acquired a two-seater Cavalier P-51 Mustang, but sadly he was killed in this aircraft when it crashed in Germany in August of that year. As a result, his collection of aircraft, one of the first private collections in the country, was dispersed. Other collections were to move in, however, as detailed in later pages of this book.

Another feature of Duxford over the years has been the large auctions of vintage aircraft held by Christies and Brooks, the well-known auctioneers. Duxford has also been used for aerobatic competitions, air racing, a staging point on air rallies and, increasingly, as a base for motion picture work. In the last two or three years, such films and TV programmes as *Memphis Belle, A Piece of Cake* and *Perfect Hero* have had large portions of aerial action filmed there.

For the future, plans include the construction of a major building to house the proposed American Air Museum in Britain, which will feature many of the Imperial War Museum's American aviation exhibits.

So, as will be seen in the rest of this book, Duxford, with its long and illustrious history, is still very much alive and a centre for vintage aircraft activity. Walking around the airfield now, it is interesting to note the number of based aircraft types that have past links with Duxford, from the very early RE8 up to the Hunter and Javelin jets via such differing types as the Cierva Autogiro, Spitfire, Hurricane, Mustang, P-47 Thunderbolt and Blenheim.

After all these years, Duxford still echoes to the sounds of the past, and long may it continue. Close your eyes and imagine the history of aviation — open them and your dreams come true!

IMPERIAL WAR
MUSEUM COLLECTION

An Act of Parliament as long ago as 1920 resulted in the creation of The Imperial War Museum, its aim being to provide a source of education and information on 20th Century military operations which had involved British and Commonwealth armed forces. To do this, it set about collecting, restoring and displaying material and artifacts relating to these conflicts . . .

The Imperial War Museum was established in the grandiose surroundings of the building previously occupied by the Bethlehem Royal Hospital in Lambeth, South East London, and these premises have been used ever since, recently receiving refurbishment. Other, dispersed, sites operated by the Imperial War Museum include the Cabinet War Room in King Charles Street, London, and the battleship *HMS Belfast,* moored on the River Thames close to London Bridge. Then, of course, there is Duxford airfield.

Space for displaying aircraft, fighting vehicles and other large wartime relics at Lambeth is at a premium and only a small selection can be exhibited there. As related earlier, the Imperial War Museum set up a storage site at historic Duxford and this later became a full museum facility in its own right, open to the public year round. The Imperial War Museum's collection of aircraft and those on loan to it are mostly non-fliers now, but span virtually the entire spectrum of aerial warfare, forming a marvellous group of warplanes of many nations, from 'stick and string' First World War biplanes to modern day jets.

FIRST WORLD WAR AIRCRAFT

Two of the rarest aircraft in the collection are also the oldest and form the 'Dawn Patrol' diorama exhibit in Hangar 4. The aircraft, an RE8 and Bristol F2B Fighter, are arranged in a very realistic reconstruction of a scene in France, complete with ground and aircrews and their sundry equipment. Together, they are a part of a large exhibition devoted to the first large-scale conflict involving military aircraft.

Originating from the Royal Aircraft Factory at Farnborough, the RE8 at Duxford, serial F3556, was built by Daimler Limited too late to see active service in the First World War. Instead, it was an early exhibit at the new Imperial War Museum, but also saw a period of display at Crystal Palace. Never a particularly popular aircraft with its crews, the RE8 or Reconnaissance Experimental 8, was used mainly for reconnaissance and artillery spotting. The RAF4a engine could power the aircraft along at up to 102 mph and it was armed with one Vickers gun firing forward and one or two Lewis guns to the rear. F3556 moved to Duxford for restoration in 1974 and has remained there since.

The RE8's successor was the Bristol F2B Fighter, or 'Brisfit', and it was used in both the reconnaissance and fighter roles with a crew of two, pilot and observer. It was armed with a forward-firing Vickers and, like the RE8, one or two aft-facing Lewis guns. Ultimately, the type was used for a very wide variety of roles and some later examples soldiered on in the Army Co-operation role until 1931. Duxford's E2581 was powered by a 275 hp Rolls-Royce Falcon III and was built by The British and Colonial Aeroplane Co Ltd at Filton as one of a batch of 500. It went on to fly with a

variety of units before passing to the Imperial War Museum at Lambeth and, later, Duxford, where it is now preserved in its original colours.

From the same era comes the Spad S.XIII biplane. This type is represented in the museum by American registered Spad S.VII N4727V, which is painted to represent an S.XIII variant, serial S.4523. This aircraft has been on loan to the Imperial War Museum for some while but originally came from Orlando in Florida.

THE INTER-WAR YEARS

At the present time, the collection does not possess examples of the splendid inter-war Royal Air Force biplanes that characterised the Duxford of this period, although the regular air displays are occasionally graced by such aircraft visiting from other collections. There is, however, one unusual aircraft that originates from this period — the Cierva C.30A Autogiro. Originally built by Don Juan de la Cierva, it was one of sixty-six production aircraft of the type. It was powered by a 140 hp Armstrong Siddeley Genet Major 1A engine. Registered G-ACUU, it was built with construction number 726 and was initially used by Air Service Training Ltd at Hamble until impressed for use by the Royal Air Force as HM580 in September 1942 after a period of storage. It was then operated by 1448 Flight, later 529 Squadron, having been rebuilt at, appropriately, Duxford, for use on calibration duties. It was later sold to the Cierva Autogiro Company, briefly becoming G-AIXE before reverting back to G-ACUU. For many years after this, it was based at Elmdon before passing to the Skyfame Aircraft Museum at Staverton, Cheltenham. Later, it returned to its old home at Duxford and in the recent past, it has been returned to the colours it wore in wartime as HM580.

THE SECOND WORLD WAR PERIOD

There are a good number of aircraft from the Second World War period in the Imperial War Museum collection, and prominent amongst these are the aircraft from the former Skyfame collection which moved to Duxford in 1978, initially on loan. They include several types of transport and training aircraft, namely the Miles Magister, Avro Anson, Airspeed Oxford and Percival Proctor, all now restored in wartime camouflage markings.

The wooden Miles M14 Magister is a tandem position, two-seater trainer built in quantity from 1937 to 1941 with other examples being built under licence in Turkey. It was the first monoplane trainer to see service in the Royal Air Force. In civilian life, it was known as the Hawk Trainer Mk III. The Magister at Duxford flew pre-war with 8 Elementary and Reserve Flying Training School at Woodley in camouflage markings with the fuselage code 'A' and civilian registration G-AFBS.

Later, it became BB661 in the Royal Air Force with whom it passed through various units before, at the end of the war, it was sold to British Overseas Airways Corporation's flying club as G-AKKU, later being re-registered G-AFBS when this earlier identity was rediscovered. A succession of owners followed before it eventually settled into retirement at

Staverton and later at Duxford. Today, it is painted as it appeared in 1940 as G-AFBS at 8 E & RFTS with camouflage fuselage and upper surfaces and 'trainer' yellow undersides and complete with blind flying hood over the rear cockpit.

Also of wooden construction is the Percival Proctor III, LZ766/G-ALCK, a de Havilland Gipsy Queen powered communications aircraft built by F. Hills and Sons in Manchester. Post-war, it flew as G-ALCK, having been registered as such in June 1948, and it is today one of few surviving examples of its type, albeit a non-airworthy one. It too wears camouflage and its original military serial LZ766.

There are two twin-engined communications types in the Skyfame part of the Imperial War Museum's collection. The first is Airspeed Oxford I, V3388. Much wartime aircraft production was shared around amongst various manufacturers and thus this particular Airspeed Oxford was in fact built by de Havilland's as part of a batch of 500 aircraft. It did not see Royal Air Force service but instead was used as a communications aircraft by the Boulton-Paul company, producers of the Defiant night fighter, eventually becoming G-AHTW on the civilian register post-war. This elegant aeroplane, powered by 375 hp Armstrong Siddeley Cheetah X seven-cylinder air-cooled radials last had a Certificate of Airworthiness in 1960 and has not flown for many years.

The other wartime 'twin' is the Avro Anson I, N4877/G-AMDA, an early example of the type built by Avro at Newton Heath with, like the Oxford, two Cheetah powerplants. It flew with a number of units involved in the delivery of wartime aircraft around the country, mainly as a crew ferry. In 1950, it became G-AMDA with the Air Navigation and Trading Company Ltd, later flying with Derby Aviation and the London School of Flying before joining Skyfame at Staverton in 1963. It flew resplendent in 206 Squadron markings as VX-F at many airshows until being damaged in a landing accident at Staverton on 2 November 1972. Since then, it has not taken to the air again although the damage has been repaired. It still remains in Royal Air Force colours.

Two more 'warlike' aircraft make up the remainder of the Second World War types from the former Skyfame collection. Z2033 is a Fairey Firefly Mk 1, construction number F.5607 from the original batch of 200 aircraft that included the four prototypes. The Firefly was a Royal Navy carrier-borne aircraft powered by a 1990 hp Rolls-Royce Griffon engine capable of propelling the aircraft at a maximum speed of 316 mph and it could be armed with cannon, rockets or bombs. Examples of the type were involved in attacks on the German battleship *Tirpitz* in July 1944, as well as operations in the East Indies against the Japanese. Later marks of Firefly

IWM's Fairey Firefly *on display in the superhangar.*

served in the East Indies well into the post-war period and one remains in airworthy condition with the present day Royal Navy Historic Flight at Yeovilton. Z2033 was converted to a target tug and re-designated as a Mk TT1. It flew in Sweden with Svensk Flygstjanst AB of Stockholm, a company well known for operating a variety of ex-military types such as the Douglas Skyraider and Gloster Meteor. The Firefly was donated to Skyfame in 1964 and subsequently camouflage replaced its target tug yellow colour scheme. The Swedish identity SE-BRD was changed for the British registration G-ASTL whilst it remained airworthy, although in retirement it has reverted to its old military identity Z2033. It has recently been restored and repainted.

The final ex-Staverton aircraft is de Havilland Mosquito, TA719, the subject of a lengthy rebuild that saw completion in 1991. Originally built in mid-1945 at Hatfield as a B.35, TA719 did not see active service but instead passed between various Maintenance Units until 1953 when it was converted to a TT.35 target tug, a task it had in common with Duxford's Firefly described earlier. After completion in early 1954, TA719 operated with 3 Civilian Anti-Aircraft Co-operation Unit from Llandow and, later, Exeter. Several surviving Mosquitos came from this source when they were declared surplus to requirements. Withdrawn in 1961, it was sold to Peter Thomas of Skyfame via 27 MU at Shawbury in 1963 and it appeared in the famous film *633 Squadron*, for which purpose it was registered G-ASKC for the flying sequences although it appeared in newly applied camouflage as HJ898/HT-G. It continued to fly with Skyfame until it suffered a landing accident on 27 July 1964, in which it was badly damaged. The port wing was written off whilst propellers, engines and fuselage all sustained extensive damage.

Later used to simulate a crashing Mosquito in that other great motion picture, *Mosquito Squadron,* further damage was caused at Boreham Wood studios. In early 1978 after a period back at Staverton painted with fuselage codes SY-G, the poor aircraft was transported to Duxford and, since then, has undergone a long period of restoration involving an almost total rebuild to many parts of the airframe. Available space in this book cannot do justice to the work involved!

The main Imperial War Museum collection also includes a variety of Second World War types, both British and foreign. The British aircraft are for the most part restoration projects at the present time with the exception of two, these being the Fairey Swordfish and Supermarine Spitfire.

The Swordfish was for many years one of the exhibits at the Lambeth site and arrived at Duxford in 1980 whilst the refurbishment was carried out on the premises there. So far, it has remained on show at Duxford. Serialled NF370, it is a Swordfish II built by Blackburns and was previously stored at Stretton before joining the Imperial War Museum.

The Spitfire is one of the Museum's most recent acquisitions, having arrived from Hong Kong in 1989. As VN485, a Mk F.24, it had flown with the Hong Kong Auxiliary Air Force and later spent a period of display in Hong Kong before coming to Britain. It looks refreshingly different to most other preserved Spitfires in its post-war silver colour scheme.

The four restoration projects come in single and multi-engined format! G-LIZY, the appropriately registered Westland Lysander III, is being restored by The Aircraft Restoration Company on behalf of the Imperial War Museum. It was built originally as V9300 and served with the Royal Canadian Air Force as 1558, later carrying the serial Y1351. It was imported to Britain in 1983 as a collection of parts by the then British Aerial Museum and until recently remained in storage at Duxford.

The other 'single' is the Hawker Sea Hurricane I, Z7015, for many years a 'gate guardian' at the Shuttleworth Trust at Old Warden aerodrome in Bedfordshire. The only surviving Sea Hurricane, this aircraft served in the Royal Navy before becoming a ground instructional airframe at Loughborough Technical College. It is under restoration to flying condition in one of the smaller buildings to the north of the main hangar complex. It has been registered G-BKTH in anticipation of its first flight.

'Heavier metal' is under restoration in the form of an Avro Lancaster and Short Sunderland flying boat. The Lancaster, a Mk X, arrived at Duxford in May 1986 after a period of ownership by Doug Arnold's Warbirds of Great Britain organisation at Bitteswell. Prior to that, it had spent some time at the Age of Flight Museum at Niagara Falls. It originally served with the Royal Canadian Air Force as KB889 and was built in Canada by Victory Aircraft at Malton, Ontario. Used by 428 (Ghost) Squadron in England, it survived active service to return to Canada and subsequent use with 107 Maritime Reconnaissance Unit. Whilst with Doug Arnold, it had been intended to rebuild the Lancaster to airworthy condition and it was allocated the civil identity G-LANC in readiness for this. At Duxford, however, it is being restored as a static exhibit and in 1990, its former RCAF colours began to be replaced by Bomber Command camouflage.

The massive (for its day) Sunderland flying boat is one of the longest running projects at Duxford. An MR5 variant, it was built by Short Brothers at their factory beside the Medway at Rochester in Kent as ML796

and it served with 228 and 230 Squadrons of the RAF's Coastal Command as well as with 4 Operational Training Unit. It was one of many transport aircraft that flew in the Berlin Airlift during 1948, operating in and out of Lake Havel. After disposal by the RAF, it went on to fly with the French Aeronavale until, in 1965, it again became surplus to requirements. It ended up in La Baule being used as a nightclub! It was saved from an ignominious end by the Imperial War Museum and was delivered to Duxford by road on trailers, the fuselage having been split into two halves horizontally from nose to tail. A major milestone in the project was achieved when the wings were reunited with the fuselage. The work on this aeroplane is ongoing and to the usual high standard expected at Duxford.

The Second World War types operated by overseas air forces and preserved at Duxford are also a varied selection ranging from the diminutive, single-seat Focke Achgelis autorotative kite to the mighty Boeing Superfortress, from restoration 'basket case' to complete exhibit in pristine condition.

The Focke Achgelis Fa330 Bachsteltze (water wagtail) was one of at least eight that arrived in Britain late in the Second World War. It was built at Hoykenkamp near Bremen by Weser Flugzeugbau with the identity 100143. Originally designed to be towed behind a submarine for airborne observation, some of the captured examples were evaluated in this country by towing them behind road vehicles! The Duxford Fa330 came from the RAE at Farnborough.

An entirely different Luftwaffe aircraft is the Messerschmitt Me 163B-1a Komet, the rocket-powered fighter which became as notorious for its accident and pilot fatality rate as it did for its dramatic climbs to height and lengthy glides back to earth. Unfortunately, there is an element of doubt as to the true identity of the Imperial War Museum's Komet. It is serialled 191660 but may actually be 191400. What is known is that it resided for some time at RAF Cranwell before being exhibited at the Imperial War Museum at Lambeth in the early 1960s. It moved to Duxford in 1976.

Also of German ancestry and with a slight doubt about its past is the Junkers Ju 52/3m 'Tante Ju'. This tri-motored, corrugated fuselage transport design was originally built by the Germans and production continued after their occupation of France in the old Amiot factory at Colombes, near Paris. Later still, the French themselves built the design under the new designation AAC.1 Toucan. The French Air Force — Armee de l'Air — flew them extensively and later, in 1960, some thirteen Toucans were transferred to the Portuguese Air Force to join some German manufactured examples. The Duxford aircraft is ex-Portuguese serial 6316 and it is believed that it is one of the French built Toucans rather than an original Ju 52/3m. Whatever its origin, it is now painted to represent a Luftwaffe aircraft with the codes IZ+NK and has often been used for ground-based film work, lending an air of authenticity to wartime films.

Many surviving Second World War types have lasted into the late 20th century having served diverse post-war careers in roles far from their designer's original intentions. Such a use was that of fighting forest fires.

An aircraft that was widely used for that purpose until safety dictated that only multi-engined types be used was the Grumman TBM Avenger, originally built as a torpedo bomber. The Imperial War Museum acquired an ex-Canadian fire fighter in 1976 and it has seen gradual preservation progress since then. This Wright Cyclone powered heavyweight originally flew with the US Navy as Bu.69327 before becoming 326 with the Royal Canadian Navy and then CF-KCG on the Canadian civil register. It was owned by Skyway Air Services in British Columbia and then another well-known BC company, Conair. Upon arrival at Duxford, it still bore Conair's red and white colours.

Ex-firefighter Grumman **TBM Avenger,** *now safely on display.*

The Republic P-47 Thunderbolt or 'Jug' was another big, single-engined aircraft. In addition to The Fighter Collection's flying example described elsewhere in this book, Duxford also has a static restoration project on the go, in fact a total rebuild using parts left over from The Fighter Collection's own aircraft. It was originally 45-49192 with the USAAF and then became FAP119 with the Peruvian Air Force. Much later, it was imported back to the USA becoming N47DD — marks now carried by Stephen Grey's 'No Guts, No Glory' — and was badly damaged in a crash at Tulsa in February 1980. It was a wrecked aircraft that commenced rebuild at Duxford in the late 1980s!

No large collection of preserved aircraft would be complete without an example of the ubiquitous Douglas C-47/DC-3/Dakota. Duxford's 'Dak' started life as constructor's number 19975 at Long Beach, California and served with the USAAF as 43-15509 before going on to fly with Scandinavian Airlines System as SE-BBH. It then flew with a number of different American owners as N9985F and N51V before once more becoming a military aircraft, this time with the Spanish Air Force with serial T3-29. Aviation film company Aces High then bought her as G-BHUB and she became well known to TV viewers as Ruskin Air Services'

G-AGIV in the serial *Airline*. The Imperial War Museum returned her to USAAF markings as 315509/WZ-S.

One of the more famous twin-engined bombers of World War Two was the North American B-25 Mitchell. Many still remain airworthy in North America following use as executive transports and fire fighters and of these, the majority owe their continued existence to the film *Catch 22* in which many of them flew, having been assembled from all over the United States. Ironically, however, the two Duxford based B-25s were not involved! The Fighter Collection's Mitchell is described in another section of this book. The Imperial War Museum's N7614C is a B-25J and flew originally with the USAAF as 44-31171 but, in the early 1960s, it flew with RCA, the American radio company. Later, it was operated as a camera ship for air-to-air photography and came to the United Kingdom in 1970 for a contract to film the first of BOAC's recently delivered Boeing 747s. It spent various amounts of time parked or impounded at Luton, Prestwick and Dublin before suddenly appearing at Shoreham in Sussex. There it was abandoned by its crew, who vanished without further formality. Finally, many months later when it was obvious that the aircraft would not be collected or outstanding dues paid, the by now decrepit B-25 was donated to the Imperial War Museum by the airport authorities. It was dismantled on site and roaded to Duxford, where extensive restoration has proceeded over a number of years.

Duxford is well known as the operating base of *'Sally B'*, the famous Boeing B-17 Flying Fortress that has flown at air displays throughout Europe for many years now. However, Duxford is also home for another B-17, a superb static restoration project known as *'Mary Alice'*. In a varied career, this Fortress has flown as a USAAF bomber serial 44-83735 and as an Oakland, California based executive transport N68629. It then went on to fly with the American religious organisation Assemblies of God Inc. It was then one of several B-17s including *'Sally B'* that flew to France to serve as aerial mapping platforms with the Institut Geographique National

Night falls over Duxford and a lone B-17.

at Creil near Paris for which purpose it became F-BDRS. Eventually withdrawn, it came to Duxford in 1982 and since then has been painstakingly returned to military condition complete with turrets previously removed, testimony to the skills of its volunteer restoration team. It now wears the USAAF serial 231965.

Of all the aircraft that have been delivered to Duxford by air, perhaps the most momentous delivery flight was that undertaken by the crew of the Boeing B-29 Superfortress 461748, *'Hawg Wild'*. Very advanced for its day, the Superfortress entered the history books as the aircraft that dropped atomic weapons on Hiroshima and Nagasaki, thus hastening the end of the Second World War. The Royal Air Force operated the B-29 as the Washington, a stop gap between the ageing Lincoln design and the soon to enter service V-bombers. *'Hawg Wild'* was built by Boeing at Renton and it served with the 307th Bombardment Group out of Okinawa. It then became a TB-29A trainer and ended its service life at China Lake, California along with other sister aircraft in use with the Naval Weapons Center. There it was left to fall apart in the desert sun until it was donated to the Imperial War Museum. It was rebuilt to flying status, registered G-BHDK and flown across America and the Atlantic, arriving at Duxford on 2 March 1980.

POST-WAR COLLECTION

The post-war collection is, perhaps not surprisingly, larger in number as, in recent years, a greater awareness of the importance of aircraft preservation has prevailed. Probably the most unusual aircraft from this era is the Saunders-Roe SRA/1 jet-powered flying boat fighter TG263. Powered by two Metropolitan-Vickers Beryl F2/4 engines (TG263 only has one left in place), this large aircraft was capable of a maximum speed in excess of 500 mph and was designed to be armed with four Hispano cannon, two 500 lb or 1000 lb bombs and rockets. Three examples were built, of which TG263 is the only survivor, the others being lost in accidents. TG263 was the first

of the type to fly, the inaugural flight being on 16 July 1947. Test pilot Geoffrey Tyson flew an SRA/1 at the 1948 Farnborough Air Display and astounded the crowds by flying long inverted passes at high speed! At the end of its flight testing, TG263 was stored although it did appear at the 1951 Festival of Britain, flying from the River Thames in manufacturer's markings G-12-1. After this, the flying boat was presented to the Cranfield College of Aeronautics before going to Skyfame at Staverton in 1966 and then Duxford in 1978.

Propeller driven aircraft of the post-war era are represented by the Handley Page Hastings four-engined transport, a type which saw use from the late 1940s, including work during the Berlin Airlift, until the late 1960s, the maritime reconnaissance Avro Shackleton, versions of which have defied the normal rules of obsolescence by continuing to serve the Royal Air Force in the 1990s, and the Vickers Varsity, a distant relative of the wartime Wellington bomber. The Varsity in the Imperial War Museum Collection served with a variety of Royal Air Force units before ending its days with the Central Flying School. After that, it spent some time flying with the Duxford Aviation Society as G-BEDV but is now grounded. It has retained its red and white RAF colours and serial WJ945.

Many of the Royal Air Force's post-war jets can be seen in the collection. These range from the Gloster Meteor F8, de Havilland Vampire trainer, Hawker Hunter F6 and English Electric Lightning F1 to the mighty Gloster Javelin FAW.9 in its vivid red and white livery in which it flew with the A&AEE at Boscombe Down and to the even mightier V-bombers — the Avro Vulcan B.2 and Handley Page Victor, the latter having been converted to an aerial tanker in 1965. A bomber from the same era and a very long serving type is Canberra B2, WH725.

A military example of that beautiful jet airliner, the de Havilland Comet, is presented in the shape of XK695, a Comet 2 that was flown on special signals duties from Wyton by 51 Squadron before retirement. Even gliders are represented by XN239, a Slingsby Cadet TX3 once flown by the Central Gliding School and typical of the type in which many thousands of Air Cadets experienced their first taste of gliding.

During July 1991 the Museum acquired an example of the McDonnell Douglas F-4J (UK) Phantom when ZE359 arrived by air. Complete with 74

Ex-Fleet Air Arm **Gannet** *277 in restored state.*

Squadron markings, it makes an interesting comparison with the similarly marked Lightning in the collection. This particular Phantom was one of several acquired by the RAF from the US Navy and can lay claim to service in Vietnam during its flying career.

Finally, a glimpse of what might have been can be seen in the form of the recently restored TSR 2, XR222. One of the prototype aircraft and originally identified on the production line as XO-4, this particular aircraft never flew, it being overtaken by political events which resulted in the cancellation of the TSR 2 programme. It survived storage at Weybridge, its place of birth, came to Duxford via Cranfield.

The Fleet Air Arm is represented by examples of the Hawker Sea Hawk, de Havilland Sea Venom, Sea Vampire and Sea Vixen and the Fairey Gannet whilst helicopters are also included in the form of two Westland built types, namely the Whirlwind and Wessex. Finally, the sole representative of the Army Air Corps is the Auster AOP.9 XP281.

A diverse selection of post-war foreign types are also to be seen. Three of them were once operated by the French Air Force, these being Lockheed T-33 14286, North American F-100D Super Sabre 42165 and Dassault Mystere IVA, 57. All came to Duxford having survived a period of storage at the USAF facility at Sculthorpe after being delcared surplus to requirements by the French. A type that originated in Canada was the Avro Canada CF-100 Canuck. Aircraft 18393, a Mk IVB, served with the Royal Canadian Air Force then came to England for trials. It was based with the Cranfield College of Aeronautics and then was purchased by collector Ordon Haydon-Baillie before being acquired by the Imperial War Museum upon his death. In 1975, it had been registered G-BCYK.

The distinctive double delta shape of the Saab Draken can be seen in the form of J35A Fv. 36075, once flown by the F16 Wing of the Royal Swedish Air Force. Larger aircraft are the USAF twin-engined transport Convair VT-29B, 17899, once flown out of Mildenhall with the 513th Military Airlift Wing and the enormous Boeing B-25D, 60689, an aircraft type that despite its age has frequently been in the headlines in the last few years, most recently for its part in bombing raids against Saddam Hussein's forces in the Gulf War.

An interesting captured aircraft is the ex-Argentine Air Force FMA Pucara, A-549. Found at Stanley Airport in the Falklands at the end of the 1982 conflict, it was shipped to Devonport after which it went to the Fleet Air Arm base at Yeovilton in Somerset. The journey to Yeovilton involved transportation as an underslung load from a Chinook helicopter during which damage was caused to the wing. It had been allocated the serial ZD487 for flight trials but because of the damage, these did not take place and instead, the airframe was robbed for spares. In November 1983, it was taken to Duxford for display after being dumped for a time at Boscombe Down.

In addition to the above detailed aircraft, the Imperial War Museum also exhibit a large number of fighting vehicles, armour and naval relics at Duxford which, although outside the scope of this book, are very well worth seeing.

DUXFORD AVIATION SOCIETY

A very large portion of the historic transport preservation movement in the United Kingdom is provided by volunteer labour. For many years, enthusiasts have been prepared to invest large amounts of time and effort (and often money) in order to preserve relics of the past and this is evident in the nation's rich heritage of restored vehicles, trains and boats as well as aircraft. Possibly the largest volunteer preservation group working on one site in the country, certainly as far as aviation is concerned is the Duxford Aviation Society, or DAS for short.

Originally formed in 1975 and a Corporate Friend of The Imperial War Museum, the DAS works hand in hand with the museum authorities on a day-to-day basis. DAS volunteer members are involved in the preservation, restoration and maintenance of the large Imperial War Museum aircraft collection. With such a large number of aircraft, it is inevitable that much work is required to just keep them clean and presentable to the public who pay to come and see them. All too often, other museums with fewer staff and resources are only able to display dusty and somewhat neglected airframes, the burden of upkeep being such a great one. In addition, however, most aircraft require more than periodic superficial work and many large scale restoration projects have been tackled over the years.

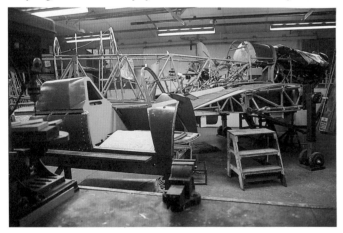

The IWM's Sea Hurricane *during restoration.*

During every weekend and on weekdays too, DAS members can be found working on their beloved aircraft. Skilled technical staff and enthusiastic amateurs work side by side in a great spirit of co-operation and this aspect of Duxford's development has been very important over the years. Without the Duxford Aviation Society, the whole collection and site would be far less developed and attractive than the one we can see today.

Duxford Aviation Society members also contribute greatly toward the upkeep of the ground facilities on the airfield and are actively involved in providing the important support services so essential on a working airfield.

These include marshalling and moving aircraft and the provision of refuelling and emergency services. Air displays at Duxford call for massive effort on the part of all concerned and DAS members come to the fore here, actually organising their own airshow every other September, intervening years being the responsibility of the IWM itself. The DAS are, however, actively involved in all of the other shows held on the site throughout the year.

The other main aim and achievement of the Duxford Aviation Society has been to acquire and preserve historic aircraft and vehicles for static display at Duxford, these being in addition to the main Imperial War Museum collection. From small beginnings, the Society and its members have worked extremely hard to assemble a superb collection of post-war British commercial airliners and transport aircraft. For many years, a small but enthusiastic band of commercial airliner fanatics called for a national policy to preserve such aircraft, but to no avail. Fortunately, the Duxford Aviation Society have fulfilled their hopes in this direction, and the results are nothing short of magnficent. Other collections in the country, particularly the Cosford Aerospace Museum and the Science Museum complex at Wroughton, have numbers of airliners on display, but those at Duxford really do form one of the finest collections in the world at the present time. Whilst there are at present some notable gaps in the collection, the efforts of the society in pursuing their acquisition and preservation policy really have to be admired.

One advantage of a preserved airliner is that its original purpose and size make it ideal for visitors to view the interior. Thus, on public open days, several airliners will be open for inspection. In this way, comparisons can be made between the various types in the collection and a dimension of interest is available that would not be possible with the majority of military types on the airfield.

Most of the airliners in the Duxford Aviation Society collection are fully restored, with two in the course of major rebuild at the present time. One refreshing aspect of the collection is that the aircraft have been painted in a variety of colour schemes that are not only pertinent to the history of each particular aircraft but also cover a spread of past airline liveries, not necessarily just the final colours carried when retired from active passenger carrying service.

For many visitors, pride of place in the airliner collection must go to the Anglo-French BAC/Aerospatiale Concorde. Concorde has proved to be the only successful supersonic passenger aircraft so far but even so, had a very long and painful development period before gaining acceptance. Built in only very small numbers, the type nonetheless represents the pinnacle of British air transport development in the eyes of many people and the appearance of Concorde at any airport or event never fails to generate great interest. The example at Duxford is one of the British-built examples from the British Aircraft Corporation factory at Filton, Bristol. It was first flown on 17 December 1971, and was subsequently used solely for test purposes, never seeing, nor intended for, fare-paying passenger service. Known as Concorde 101, its manufacturer's construction number was 13522. It was allocated the British civil aircraft registration G-AXDN, and this identity is still carried on the preserved aircraft, which wears its manufacturer's livery along with various 'adornments' on the forward fuselage, reminders of test assignments carried out over the years in far flung locations.

In January 1974, G-AXDN became the fastest Concorde ever when it attained the speed of 1450 mph (Mach 2.23). It also holds the record for the fastest crossing of the Atlantic between Europe and the USA by a civilian airliner, this being achieved in the incredible time of two hours fifty-six minutes. After being used for training at Fairford in Gloucestershire for some time, the aircraft became surplus to requirements and was obtained by the Duxford Aviation Society in 1975. It was eventually flown into Duxford from Filton on 20 August 1977 by Brian Trubshaw, who had flown the original British prototype on its first flight. Externally, Concorde 101 is representative of operational examples with the exception of its length. Later aircraft in airline service had a rear fuselage extension beyond the fin, which added a further ten feet in length. It is one of two Concordes on public display in the United Kingdom, the other being preserved at the Fleet Air Arm Museum at Yeovilton in Somerset.

At the other end of the post-war airliner scale in terms of speed, comfort and looks is the collection's Avro York. The type was originally conceived as a military transport, the prototype first flying in July 1942. It made extensive use of Avro Lancaster bomber components, namely the wings, engines and tail, but the fuselage was of a new design. In the immediate post-war era, Yorks flew on passenger routes all over the world, as well as contributing much toward the efforts of the Berlin Air Lift. It was used by airlines such as British Overseas Airways Corporation, South African Airways, British South American Airways Corporation and the Argentinian airline FAMA. Later, many British independent airlines operated the type and it was used extensively for trooping contracts on behalf of the British Government. In all, 265 Yorks were produced by Avro — a further solitary example was built in Canada — and of these, some eighty-four appeared on the British Civil Aircraft Register, most of which were ex-RAF machines.

Duxford's York, G-ANTK, was formerly serial MW232 in the Royal Air Force, with which it flew on the strength of 242 and 511 Squadrons. It was operated by Dan-Air from the autumn of 1956 until being retired eight years later. Subsequently, it spent many years in use at Lasham in Hampshire as a bunk house for Air Scouts on gliding courses! Later, it became part of the Dan-Air Preservation Group until this disbanded in 1982. Dan-Air eventually presented the York to the Duxford Aviation Society collection and it arrived at Duxford by road on 23 May 1986. It has since been undergoing a major rebuild inside one of Duxford's old RAF hangars.

Dan-Air has donated two other aircraft to the DAS collection, Airspeed Ambassador G-ALZO and de Havilland Comet 4 G-APDB. The Airspeed AS57 Ambassador was a sleek, twin-engined, triple-finned airliner powered by two Bristol Centaurus radial engines. It was originally meant to be a Dakota replacement but the design evolved so that it eventually saw the light of day as a medium-haul, high speed airliner. All twenty-one production aircraft flew for British European Airways, who named the type the 'Elizabethan'. G-ALZO was built at Christchurch as constructor's number 5226 and flew with BEA as 'Christopher Marlowe'. In 1960, it was sold, along with G-ALZP and G-ALZY, to the Royal Jordanian Air Force with which it was serialled 108. It was used as a VIP transport. In early 1963 it was restored to the British register by Dan-Air. In company with others of the type operated by Dan-Air, it flew from several British airports, always a stirring sight on start-up when, more often than not, the Centaurus powerplants would belch thick clouds of smoke!

Taken out of service in September 1971, G-ALZO remained at Lasham and was cared for by the aforementioned Dan-Air Preservation Group before being donated to the DAS. After arrival at Duxford in October 1986, work commenced on its long-term rebuild.

The third Dan-Air airliner at Duxford, Comet 4 G-APDB, is a nicely preserved example of this elegant aircraft. The Comet 4 was developed from earlier models after the tragic loss of several examples through accidents later found to be caused by catastrophic structural failure. By the time that the redeveloped aircraft entered service in 1958, the British lead in jet powered commercial aircraft had effectively been taken by the Americans, notably in the form of Boeing's 707 and the Douglas DC-8.

G-APDB, constructor's number 6403, first flew on 27 August 1958, from its birthplace at Hatfield and it subsequently enjoyed a lengthy career, initially with BOAC with whom it served for six years before being sold to Malaysian Airways as 9M-AOB. Dan-Air acquired it in mid-1970. It took on the British registration G-APDB once more and was finally grounded in 1973. Its final flight was from Lasham to Duxford on 12 February 1974, and it remains resplendent in its Dan-Air colours of red and white. Whilst with BOAC, G-APDB gained the distinction of being the first jet aircraft to fly across the Atlantic in an easterly direction carrying fare-paying passengers. It is interesting to note that sister aircraft G-APDC was making history simultaneously by flying the same trip westbound.

One of Britain's most successful commercial aircraft has without doubt been the Vickers Viscount and in the 1990s a number are still in revenue earning service. The last of 444 examples rolled off the production line in 1964 and the type has seen extensive civil and military use the world over. The model 701 Viscount was powered by four Rolls-Royce Dart R.Da.3 Mk 506 turboprops and had a cruising speed of 316 mph and a range of 1450 miles. Its standard of comfort and large oval windows made it a popular aircraft with passengers. The Viscount 701 at Duxford is G-ALWF, the fifth Viscount to be built. It flew for the first time on 3 December 1952, and was delivered to BEA as 'Sir John Franklin'. Later, it flew with Channel Airways and British Eagle before being sold on to Cambrian Airways in January 1966. Withdrawn from use in December 1971 it was preserved at Liverpool's Speke Airport for a time before being roaded to Duxford on a large articulated trailer. It is now resplendent once more in the original maroon grey and white livery of BEA — pure nostalgia for the airliner buff!

Bristol's 'Whispering Giant', the Britannia, is represented in the collection by G-AOVT, a Series 312 aircraft powered by four Bristol Proteus powerplants which gave the type its nickname. Built with the line number 13427, it entered service with BOAC after its first flight in December 1958. Sold to British Eagle in 1963, it passed on to the Luton-based Monarch Airlines in May 1969, after the collapse of its former owners. It was retired at Luton in March 1975, and flew to Duxford three months later. It has remained in Monarch's striking yellow and black colour scheme since that time.

Another aircraft in the collection with Rolls-Royce Dart turboprops is the Handley Page HPR7 Herald, in fact a Series 201 model registered G-APWJ, constructor's number 158. Like the Airspeed Ambassador before it, the Herald was originally conceived as a Dakota 'replacement'. The prototype was actually powered by four Alvis Leonides radials but production aircraft switched to the popular Dart. Twenty-three 'Dart Heralds' were built and a small number are still flying in the early 1990s.

The Herald at Duxford was built in 1963 for British United Airways and flew in their attractive white, turquoise and gold livery for many years, later passing on to British Island Airways Ltd in May 1970, and then to Air UK. It flew into Duxford from its former base at Norwich on 7 July 1985, and still retains Air UK's colours.

Dominating the skyline at the eastern end of the airfield is the tailplane of the DAS Vickers Super VC-10. The VC-10 was built to serve on BOAC's long-haul routes around the world at high speeds and weights. With its high fin and swept tail, it was a most attractive and popular airliner. Powered by four rear-mounted Rolls-Royce Conway jets, the type 1151 was capable of a maximum cruise speed of 580 mph, a range of 5960 nautical miles and a ceiling of 38,000 ft. Main customer for the VC-10 was BOAC, although the type did sell in smaller numbers to other airlines both at home and abroad. Additionally, fourteen were ordered by the Royal Air Force as long-range transports and all except one remain in service today, the odd one out ending its days as a test bed for the Rolls-Royce RB211 engine. Later, the RAF purchased surplus British Airways (as BOAC had become) and other foreign VC-10s for conversion to the air-to-air refuelling role and so the type remains very much operational in this important strategic role.

The VC-10 at Duxford is G-ASGC, which started life on the Vickers production line as constructor's number 853. It spent its entire active life with BOAC/British Airways and after retirement, was ferried to Duxford, the final flight taking place on 15 April 1980. It was subsequently stripped of its BA colours, these being replaced by the splendid BOAC Cunard colours of white, deep blue and gold that it wears now.

British Airways were also the source of another airliner in the DAS collection, namely the Trident 2E G-AVFB. This was originally built for British European Airways, the other partner, with BOAC, in what became BA. The model 2E featured an increased wing span compared with earlier versions and was also built with fuel tanks in the fin and more powerful Rolls-Royce Spey powerplants. BEA originally ordered fifteen examples of the Trident 2E and were indeed the major customer for all marks of Trident. Distinguished from other British jet airliners by its three rear-mounted engines, the Trident's success was overshadowed by the American Boeing 727 family which featured a very similar appearance but was built in far greater number. In 1990, Trident G-AVFB entered the Duxford 'Superhangar' in its BA livery and later emerged resplendent in the old BEA colours of black, white and red, so much a part of the Heathrow Airport scene in the 1960s.

The remaining aircraft in the DAS collection are something of a miscellany! The fuselage of Handley Page Hermes IV, G-ALDG, was used for many years as a stewardess training facility at Gatwick Airport and was a familiar sight to passengers travelling along the adjacent London-Brighton railway line, initially in British United Airways colours and later in those of British Caledonian Airways. It has flown with several airlines during its flying career, namely BOAC, Falcon Airways, Britavia and Silver City. It was dismantled at Gatwick in October, 1962, and began its second, ground-based role before eventually being acquired by the DAS in January 1981. It is now painted in BOAC colours and although incomplete — it still lacks wings — makes an interesting comparison with the Imperial War Museum's Handley Page Hastings, the design of which was closely related.

Before the days of executive jets, the de Havilland Dove saw use as a corporate transport aircraft as well as a mini-airliner. A small number were used by the Civil Aviation Flying Unit at Stansted. G-ALFU, a Dove 6, was one of these aircraft and flew on calibration work for airfield radar systems throughout the United Kingdom. It came to Duxford originally for the Imperial War Museum but later passed to the Duxford Aviation Society who have cared for it ever since, despite some unfortunate gale damage sustained in the late 1980s, now fortunately repaired.

Finally, the upper nose section of Blackburn Beverly XB261 was acquired in May 1989 when this aircraft met a most ignominious end at the hands of a scrap merchant at Southend the previous month. The aircraft had originally been saved from such a fate and was exhibited at the Southend Historic Aircraft Museum until this closed. Thereafter, the unfortunate 'Bev' lingered on until its appointment with the blowtorch.

THE FIGHTER COLLECTION

By far the largest privately owned collection of 'Warbirds' in Europe and rivalled by very few anywhere in the world, The Fighter Collection started in early 1980, when Stephen Grey obtained the first of the Collection's Second World War fighter aircraft, a North American P-51D Mustang. Since then, the collection has steadily expanded to encompass the cream of British and American fighters from the 1939–1945 era. During the airshow season at Duxford, The Fighter Collection line-up covers a very large portion of the flightline. When not flying, Europe's finest active warbird squadron fills an entire T2 hangar toward the eastern end of the airfield.

The Fighter Collection first moved to Duxford in April 1984, when the Mustang, Grumman Bearcat and Wildcat flew in from their original base at Geneva in Switzerland. All three aircraft had previously appeared at a select few shows in England. The Wildcat has since passed on in a trade with an American collection, but the other two aircraft remain as founder members. Duxford has been the operating base of The Fighter Collection ever since, and a major attraction for any visitor, whether on an airshow day or not. The Collection epitomises all that is good about Duxford — rare aircraft that are alive and flyable, restoration projects that can be viewed close to, and a band of dedicated, enthusiastic and talented engineers, some full-time, others volunteers.

The concept behind The Fighter Collection can be divided up into several aims and objectives. Primarily, the aim is to acquire examples of the leading combat aircraft types from the Second World War period and, to date, the main effort has been expended, naturally enough, on obtaining fighters. In collecting these aircraft, the object is either to keep existing airworthy aircraft flying and to return them to their original standards or, in the case of non-airworthy machines, to restore them to original flying condition.

The Collection considers it most important that their aircraft are seen regularly in their natural element, so that enthusiasts today, and those to come, can appreciate the power and beauty of these fighting machines. To achieve these aims, The Fighter Collection is also dedicated to bringing together as much support for their operation as possible. This support not only includes the highly skilled group of air and ground crews and their joint experience and expertise, but also the historical and technical documentation to enable the aircraft to be maintained and displayed in flying trim and in historically accurate colour schemes.

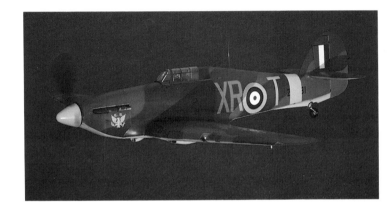

Right *Cloud storming. TFC's beautiful* **Hurricane XII**, *in the capable hands of Nick Grey, slips through the Cambridge skies that fifty years earlier this Hawker classic defended during the height of the Battle of Britain.*

Right *Home to 'The Few'. Simple but efficient.*

Finally, as the years since the last world conflict rapidly pass by, the Collection seeks to maintain sufficient spares holdings or manufacturing capability to ensure that continued airworthiness is maintained for the foreseeable future.

The Fighter Collection's aircraft are always a major feature at Duxford's air displays but they also travel widely throughout the United Kingdom and far into Europe, where they are seen at many shows, large and small. Frequently, a large portion of the Collection fleet will fly out of Duxford and spend a weekend away, thrilling audiences with their performance.

In the last few years, the Collection's aircraft and pilots have been heavily involved in aviation filming as well as air display flying, the aircrews using their wealth of expertise and experience for this demanding type of work. The aircraft have provided the all-important authenticity in such films as Spieleberg's *Empire of the Sun,* David Puttnam's *Memphis Belle* and television programmes such as the controversial Battle of Britain-based *Piece of Cake, Christabel* and *A Perfect Hero.*

Numerically, the main emphasis of the Collection is upon single-seat American fighter aircraft, and it includes examples of the finest of the breed to be found anywhere in the world. American types in the Collection include the North American P-51 Mustang, Grumman Bearcat and Hellcat, Republic P-47 Thunderbolt, Chance Vought Corsair, Curtiss P-40 Kittyhawk, P-63 Kingcobra and Lockheed P-38 Lightning. Seen together, these aircraft exude speed and power, muscle and elegance, a tangible link with a past where pilots relied on their skill without the benefit of late 20th century computer wizardry and fly-by-wire technology. In their day, they were all 'state of the art' fighting machines.

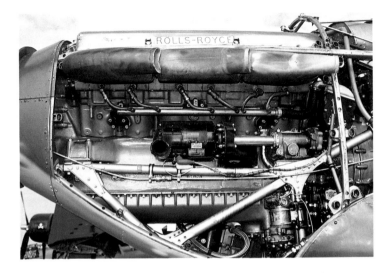

Above *Rolls-Royce magic. The Merlin.*

Previous page *Duxford beckons. Carl Scholfield unleashes the mighty Merlin in a final salute to the last Great Warbirds Air Display and to historic West Malling, an airfield which bore the brunt of Goering's Luftwaffe in the summer of 1940.*

Opposite *Hurricane at rest. A silent tribute to the sacrifice of the few for the many.*

HAWKER HURRICANE XII

For many years, the Spitfire has tended to overshadow its fighting partner in the Battle of Britain — the Hawker Hurricane — despite the fact that in that conflict, the Hurricane was numerically more significant. The Hurricane's cause has, perhaps, not been helped by the fact that far fewer have survived in flying condition. Until recently, the two examples operated by the Royal Air Force Battle of Britain Memorial Flight were the only flying Hurricanes in Europe, a third example having returned to its original home of Canada some years ago. It was, therefore, highly

Above *Warrior at peace. The* **Spitfire,** *one of the most formidable fighting machines of the Second World War.*

Right *Through the vertical. Mark Hanna pulls inverted in OFMC's Mk IX as Nick Grey goes ballistic.*

significant and appropriate that The Fighter Collection managed to complete the total rebuild of their own Hurricane in time for the 1990 air display season, which saw many celebrations of the 50th anniversary of that famous air battle.

The only Hurricane currently flying in private ownership on this side of the Atlantic was built in 1942 by Canadian Car and Foundry and was originally powered by a Packard-built Merlin engine. It was identified on the production line as c/n 72036 and later became 5711 in RCAF service. It served in the provinces of Nova Scotia and Quebec until being 'mothballed' in 1945. Like many ex-RCAF aircraft, it was sold to a farmer at a knock-down price and he no doubt intended to strip the airframe of any parts that could be used around his farm. Many surviving aircraft that have come out of Canada over the years had lain virtually forgotten on farms, long since robbed of components for agricultural machinery and buildings.

Right *Fly low, fly fast. Hoof Proudfoot, TFC's chief pilot, leads Paul Bonhomme in the* Hurricane XII.

Overleaf *Hoof again. This time a superb piece of formation flying illustrates the legendary lines of Mitchell's creation.*

By the 1970s, Hurricane 5711 was to be found on a farm at Regina, Saskatchewan. It was purchased by a group of enthusiasts who intended restoring it, but ultimately the project was passed on to The Fighter Collection. It was imported to the UK in 1983 and spent periods of time at Coventry and Coningsby before being transported to Duxford, where a Fighter Collection team completed a meticulous and total rebuild.

With restoration completed, the airframe was painted as aircraft 'XR-T' of 71 Squadron, Royal Air Force. As such, it represents a Hurricane IIB, although the designation applied to the Canadian-built aircraft was in fact Hurricane XIIA. 71 Squadron was re-formed in September 1940, having earlier been a First World War Sopwith Camel unit, and was manned by American volunteer personnel, thus becoming the first so-called 'Eagle' Squadron. Their eagle insignia is carried on the port side of the engine cowling, whilst the starboard side bears the famous Disney cartoon cockerel character in boxing gloves, also carried by 71 Squadron.

First flown after its rebuild on 1 September 1989, it became a much sought-after display item at airshows during the Battle of Britain 50th anniversary year. The restoration won the *FlyPast* Magazine Mike Twite Memorial Trophy, having been judged by readers to have been the most significant contribution to aircraft preservation in the United Kingdom during 1989.

Although flown in military markings, the Hurricane is on the British Civil Aircraft Register as G-HURI, these highly appropriate markings having been allocated as far back as June 1983.

SUPERMARINE SPITFIRE LF IXE

The aircraft making up The Fighter Collection are not confined to American fighter types — both the British Spitfire and Hurricane are represented. In fact, the collection has several Spitfires on rebuild to airworthy status, but the best known is their flying example, ML417/G-BJSG. Built at the large Castle Bromwich factory during 1944 as an LF IXE variant, it was first flown operationally by 443 Royal Canadian Air Force Squadron, initially from Ford in Sussex. It carried the squadron code markings 21-T, those that it still carries today.

Whilst being operated in France in the summer of 1944, ML417 saw combat against the Luftwaffe and was involved in 'damage' claims on an Fw 190 and Bf 109. It also sustained some damage from flak itself. Later, whilst flying from a Normandy base, it was pitted against several Bf 109s, two of which were destroyed. This action took place on 29 September 1944. Until the end of the war, ML417 stayed with Canadian units, flying with 442, 401, 441, 412 and 411 Squadrons.

After a time of inactivity at a Maintenance Unit in Shropshire, Vickers Armstrong purchased the aircraft and, having converted her to a two-seater trainer, supplied ML417 to the Indian Air Force as HS543 after a brief appearance in the manufacturer's test markings G-15-11. Later, it was stored at Palem with The Indian Air Force Museum before sale to America and further inactivity in Colorado. After being acquired for The Fighter Collection, the airframe was converted back to its single-seater status.

Registered G-BJSG in January 1981, it has continued to fly as ML417 in 443 Squadron colours, although from time to time it has temporarily worn other liveries for film work, most notably in *Piece of Cake*. Although by rights a 'clip-wing' Spitfire, temporary pointed wing tips have been instal-led when necessary to more accurately represent a Battle of Britain participant.

As mentioned earlier, The Fighter Collection has several Spitfires under rebuild and, indeed, one has already been completed and has for some time operated in New Zealand. Others undergoing the TFC treatment include a Mk XIV and FR XIV.

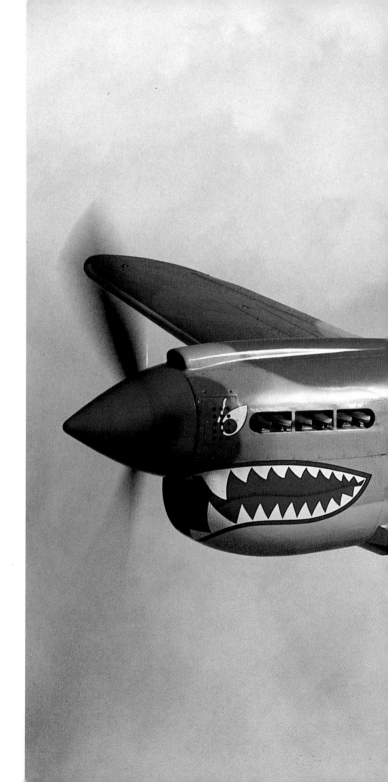

CURTISS P-40 KITTYHAWK

Not so long ago, Duxford could boast not one but two resident Curtiss Kittyhawks, one with The Fighter Collection, the other with The Old Flying Machine Company. The latter eventually returned to the United States, leaving The Fighter Collection's P-40 as yet another unique rarity in European skies.

Duxford's Kittyhawk was built as a P-40M, United States serial 43-5802, constructors number 27490, although it passed to the Royal Canadian Air Force as "840". It ended its Canadian days on Vancouver Island, British Columbia. After being categorised as surplus, it became N1233N on the United States Civil Aircraft Register and operated in the Seattle area, later spending some time at Oregon State College before passing to Columbia Airmotive, with whom it remained from the mid-1950s to 1979. The rebuilt aircraft became NL1009N with Tom Camp, who eventually passed it to The Fighter Collection in 1984. It arrived at Duxford, inside a crate, in bare metal finish in February 1985, and was carefully assembled and painted in a superb colour scheme to represent aircraft FR870 of 112 Squadron, Royal Air Force as flown during 1943 in the Western Desert. Coded GA-S, it carries the famous 'sharkmouth' insignia on the nose. This type of marking has been worn by many and varied aircraft types of numerous nationalities, the first recorded useage being as long ago as 1916, and to this day combat aircraft carry the ferocious looking decoration. The marking somehow looks 'right' on the P-40, probably because of the shape of the nose and propeller spinner, and 112 Squadron made the sharkmouth symbol its own.

GRUMMAN F6F HELLCAT

Another Grumman type in The Fighter Collection is one of the more recent acquisitions, the F6F-3 Hellcat. It arrived during 1990 by sea and was assembled in time to appear at the final Duxford airshow of that season in Royal Navy markings, but without a serial number. Originally Bu.40467, the Hellcat was acquired from Ed Maloney, was later restored and adopted the paperwork of N11TF, an F6F-5 which suffered a forced-landing at San Marcos, California, in April 1979.

Right *Shark's teeth and the P40 are an eternal pairing. This example of the P40N is painted in the 1943 Western Desert Campaign colours of 112 Squadron, RAF.*

Overleaf *Blue sky, winged death. Nick Grey bares the Kittyhawk's teeth.*

Right *Fearsome fighter. The* Kittyhawk *is hauled by a 1200 hp Allison V-12 liquid cooled engine.*

During March 1991, the Hellcat was resprayed in its original colours at Duxford, having been re-registered as G-BTCC. F6F-3 Bu.40467 is actually the aircraft in which US Navy ace Lieutenant Alex Vraciu scored seven airborne 'kills' in two sorties while he and the aircraft were serving with Navy Fighting Squadron VF-6.

GOODYEAR FG-1D CORSAIR

One of no less than four airworthy Corsairs based at Duxford, The Fighter Collection's 'bent-winged bird' is a Goodyear-built FG-1D model with three instead of four-bladed propeller and fabric covered wings. This particular aircraft actually saw combat having been delivered to the United States Navy in early 1945 as serial Bu.88297, constructor's number 3111. Initially allocated to Carrier Air Group 2 then VFB-2, it subsequently flew with CV-102 on board the *USS Aatu,* seeing action in the Mariana Islands. By 1948, it was flying from Memphis with a Naval Air Rescue unit, later seeing periods of duty in Minnestoa and Florida.

The late Frank Tallman, famed for his aviation filming activities, ensured the aircraft's future by rescuing it from the scrapman's torch in 1959, and it remained in his ownership, unconverted from military stock condition, until 1967. During this period, it was displayed in the Tallmantz Collection at Orange County, California. In his book, *The Great Planes,* Tallman details the flying characteristics of the aircraft and describes it as his all-time favourite aircraft. Registered as a civilian aircraft with the markings N9145Z, it passed to the Minnesota Aircraft Museum, Minneapolis (one of its former military bases). The Fighter Collection acquired the Corsair in

1986, it having been re-registered N8297 by this time in recognition of its earlier military serial.

This distinctive 'Bent Winged Bird' now flies in a blue Unites States Navy colour scheme, representing the Corsair flown by Lieutenant Ira Kepford of VF-17 Squadron as flown in the Pacific during 1944. It bears the white fuselage code '29' and a skull and crossbones insignia on the engine cowl. Underneath the cockpit appears a tally of Japanese "kill" markings.

GRUMMAN F8F BEARCAT

The Grumman Aircraft Company has, over the years, produced many famous naval fighting aircraft. Perhaps the ultimate single piston-engined fighter was the Grumman Bearcat. Again, a great rarity outside the United States, The Fighter Collection's example is a very popular airshow performer and has put on many dazzling displays over the years, mainly in the hands of Stephen Grey and the late Stefan Karwowski. This superb aircraft was built in 1946 as an F8F-2P variant, with the manufacturer's identity 1088 and the US Navy serial Bu.121714. It flew with various units, including VC-190, VC-61 and VU-2, being based at such locations as

Below *Awesome! Stephen Grey roars low across the grass at Duxford, leading Jack Brown in the FG-1D.*

Right *Face to face with hell. This baby could ruin your whole day, and often did for the Japanese in the Pacific theatre.*

Alameda, Quonsett Point, Cherry Point and Norfolk, Virginia. After a period of storage, it ended up at the Ontario Air Museum in California in the care of collection owner Ed Maloney, going there in 1962 and acquiring the civil registration N4995V. It made occasional forays into the air and was later sold to Bubba Beal and Chubb Smith of Knoxville, who re-registered it as N1YY. After a total rebuild it became, in turn, N700H and N700HL, and was acquired by The Fighter Collection in 1982. By that time, it had already been repainted in its present colours, those of United States Navy Squadron VF-41 — *The Red Rippers* — and wears the individual aircraft code 100/S. More recently, VF-41 have flown Grumman's modern day equivalent of the Bearcat, the F-14 Tomcat.

There is a second Bearcat in The Fighter Collection, albeit under restoration at the present time. An F8F-1B version, it was originally serialled Bu.122095 with the US Navy, having been built as constructor's number 779. Later, it served with the Thai Air Force and was displayed outside a Government building in Bangkok after its retirement. It came to Duxford in June 1988, via the famous Jean Salis collection at La Ferté Alais in France.

Right *A massive 14 ft propeller is needed to transmit the R2800's 2000 hp powerplant. Nick Grey kicks the* F6F *into life.*

Below *Stephen Grey drops the* Hellcat *over the threshold — Classic Fighter Display, 1991.*

NORTH AMERICAN P-51D MUSTANG

The first aircraft in what was to become The Fighter Collection was the North American Mustang. A P-51D-25-NA model, first registered N6340T, was acquired in May 1980. This aircraft originated on the North American production line at Inglewood, California, as constructor's number 122-39608 and it was delivered to the United States Army Air Force in England during February 1945, with the serial number 44-73149. Its combat service is not known. In mid-1947, it was transferred to the Royal Canadian Air Force, who changed the aircraft's identity to '9568'. It flew with the Station Flight at Suffield, Alberta but was surplussed in early 1957 and became N6340T on the US Civil Aircraft Register. In time, it received a second seat and radio equipment that more readily enabled it to operate as a civil aircraft. It passed through two owners in California before being purchased from Dr Robert MacFarlane. With the aid of under-wing ferry tanks and the capable hands of racing pilot John Crocker, the Mustang was flown across the Atlantic via the Northern route, stopping in Iceland and at Gatwick *en route*.

Up to this time, N6340T had flown in a civilian colour scheme of bright red with white trim, with the name *Candyman* on the nose and fuselage and the race number *7* on the fin. In this guise, it had been used once as an air racer during the mid-1970s by Charles Beck, fortunately remaining in basically 'stock' configuration without suffering the hideous modifications that

Above and Right   *Combat veteran. Meticulous attention to detail is apparent throughout The Fighter Collection. The out of true kill markings are identical to those it wore during World War II.*

have been applied to some American warbirds in the quest for added speed.

Having decided on an authentic wartime colour scheme for the new acquisition, the P-51 was repainted — at London Heathrow Airport of all places! — in the colours of the USAAF 8th AF P-51D-20NA serial 44-63221 as flown by the 362nd Fighter Squadron, 357th Fighter Group, thought to be the Group to which the original aircraft was supplied. It bears the fuselage codes G4-S, distinctive red and yellow chequers on the nose, black and white 'invasion' stripes and, in a nod to the past, it continues to carry the name *Candyman* on the starboard nose cowl, the position favoured by mechanics, who often had a different nose art to the left-hand or pilot's side. On the port side, the name of the original 44-63221 is worn, this being *Moose,* after Lieutenant 'Moose' Becroft who flew her during 1944. A cartoon moose head appears by the name, together with a group of Swastika 'kill' symbols.

It made an exciting addition to the European air display scene upon its debut, the first Mustang to be seen since the late 1960s and it has continued to thrill show crowds ever since.

In June 1985, the new American registration N51JJ was allocated and the aircraft went on to star in the films *Empire of the Sun* and *Memphis Belle.* For the former, it was repainted silver overall with brown and orange trim, with the name *My Dallas Darlin'* and tail code 583 whilst in *Memphis*

Right   *Hoof Proudfoot thunders into the still, evening air at West Malling, in the Goodyear-built* Corsair.

Below   *The skull and crossbones emblem of VF-17, one of the deadliest outfits of the Pacific War.*

Belle, it flew in USAAF green with fuselage code AJ-S but retained the *Moose/Candyman* titles. When not involved in filming, the Mustang has always been restored to its normal colours, making it one of the most attractive P-51s around.

During the winter of 1989/90, N51JJ received a major rebuild at Duxford, returning it to an almost completely 'stock' military configuration, even down to a working gunsight. In January 1991, it was re-registered with the British identity G-BTCD. This aircraft remains one of the most popular fighters in the collection, not least with her groundcrew, who no doubt value the second seat!

REPUBLIC P-47 THUNDERBOLT

Whilst, numerically, the Mustang is fairly well represented in Europe, the Republic P-47 Thunderbolt is undoubtedly a rarity. The example owned and displayed by The Fighter Collection is unique in European skies. This somewhat portly aircraft, known affectionately during World War II as 'The Jug' by its crew, does in fact exhibit rather lively flying characteristics, especially when going downhill! With its curved lines accentuated by the

Left *Fix the biggest propeller to the most powerful engine available; hang it on the front of the smallest airframe you can get away with; and you have breathtaking performance that outshone even the early jet fighters.*

Right *Cat pack.*

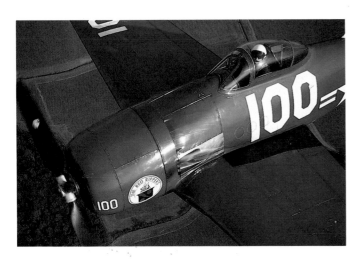

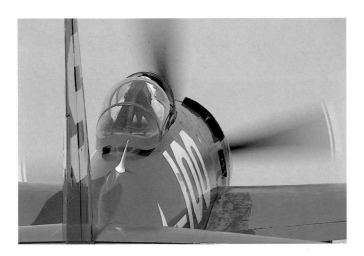

Above and Right *Too late to see action in World War II, the* Bearcat *saw considerable action in Korea.*

'teardrop' canopy rather than the razor back fuselage of earlier models, the 'T-Bolt' even *looks* attractive, despite its bulk.

The Fighter Collection's P-47 was restored by Steve Hinton's Fighter Rebuilders in Chino, California and, following test flights in the United States registered as N47DD, it was crated and shipped to Duxford, where re-assembly took place. Initially, it wore a silver scheme with red nose cowl and rudder under the marks 42-23719/ fuselage code 19. Later, the red cowl gave way to a black and white chequered finish, whilst 'invasion' stripes were applied to the wings. Now coded MX-X, it represented aircraft 42-26671 of the 82nd Fighter Squadron, 78th Fighter Group, as operated from Duxford in 1944–1945 and flown by Lieutenant Colonel Ben Mayo. It bears the aircraft's name, *No Guts, No Glory,* beneath the cockpit canopy. It is now one of very few Thunderbirds flying anywhere in the world.

NORTH AMERICAN B-25 MITCHELL

All of the aircraft so far described have been high performance, single-engined fighter aircraft. The Fighter Collection also operates a twin-engined bomber, the B-25 Mitchell, which in addition to being a display aircraft in its own right, also acts as a 'mother ship' when several of the collection's aircraft fly away from base to shows. It is a useful and exciting crew bus for the mainly volunteer ground crew who care for the aircraft in The Fighter Collection.

Once operated by the USAAF as 43–3318, c/n 100–20644, this North American B-25D-30NC was the final aircraft in a batch of twelve that was later flown by the Royal Canadian Air Force. It was delivered to the RCAF on 18 January 1945. It carried the serial KL161, and served with various units from such bases as Boundary Bay, British Columbia; Trenton; Cold Lake and Ottawa. In the early 1960s it was sold to an individual named Joe Goldney. It stayed in and around the Vancouver area for some time, registered as CF-OGQ, before spending a period in Alaska, where it became N88972.

During the mid-1980s, The Fighter Collection commissioned a rebuild by Aero Trader at Chino in California. Stephen Grey was part of the ferry crew when the aircraft was flown across the Atlantic on its delivery flight

Right *Superb visibility and a wide track undercarriage helped make the F8F an excellent carrier-borne aircraft. They also come in quite handy at Duxford in the 1990s.*

to Duxford. It arrived at its new home in November 1987, in bare metal finish.

N88972 remains the Mitchell's official identity, but she flies in the colours of 98 Squadron, Royal Air Force, with the individual aircraft codes VO-B and, on the port nose, the name *Grumpy* together with the appropriate dwarf cartoon character from the famous Walt Disney film. The starboard nose is emblazoned with The Fighter Collection's emblem, the Lafayette Escadrille Red Indian head.

BELL P-63-A-7BE KINGCOBRA

Another aircraft currently receiving the full loving care and attention of The Fighter Collection's ground crews is Bell P-63A-7BE Kingcobra, USAAF serial 42-69097. Acquired by The Fighter Collection in early 1991, the airframe is having the 'full works' and hopefully will be refitted with a new

Right   *Having earned their keep during the Classic Fighter display 1990,* Moose *and* No Guts No Glory *bask in the late afternoon sun.*

Overleaf   *Moose and Candyman, one aircraft — two nose arts, both representative of 357th Fighter Group, Cambridgeshire, 1944.*

engine and painted in time to take its place with the rest of the team during the 1992 show season.

The majority of the Kingcobras built were supplied under Lend-Lease to the Russians, who used them to good effect in the ground attack role. Only relatively few were retained for use by the USAAF, mainly in the training role.

42-69097 flew with various USAAF Base Units, namely the 432nd at Portland AFB, the 264th at Peterson AFB, the 205th at Fairmont, the 209th and 421st and finally, the 211th at Sioux Falls. On 24 October 1945, the Kingcobra passed to the RFC Disposal Base at Altus, Oklahoma, for disposal with a total of 730 hours 10 minutes flying time to its credit, having been reconditioned and redesignated as a RP-63-A a few days before.

Throughout its service with the USAAF, it appears to have suffered just one minor accident whilst taxying on 4 June 1945, when damage to the right wing and propeller was caused. The damage was not serious, however, and repairs were quickly effected with the aircraft taking to the air again on 13 June.

Whilst subsequent details are sketchy, it would appear that 42-69097 continued to fly, as by June 1945 its log shows a total of 815 hours. The aircraft was acquired by Flamingo Aircraft Inc on 28 April 1947 and in

Right *Nicknamed 'The Jug', the Republic P-47* Thunderbolt *was a huge aircraft, capable of absorbing terrific punishment. Operated by both the RAF and the Americans, one pilot is quoted as saying, 'If you got bored in a P-47, you just got up and took a walk around the cockpit.'*

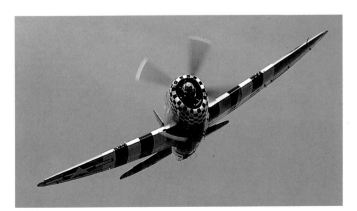

June of that year changed hands to A. T. Whiteside, with whom it was registered NX52113.

Little or no flying occurred from that date, although the aircraft was maintained until acquired by Johan M. Larsen in 1966 and exhibited in a museum at Minneapolis until 1973. The aircraft was then acquired by The Military Aircraft Restoration Company of Long Beach, California.

In the mid-1980s, it was given a 'clean bill of health' by the American authorities and, in early 1988, sold to Doug Arnold's Warbirds of Great Britain and shipped to England. Some further repair and maintenance work was carried out at Biggin Hill prior to its move to Duxford and the Fighter Collection's acquisition in early 1991.

At the time of writing, no decisions had been made to the final scheme of markings for the aircraft.

Right *Despite its size, the* Thunderbolt *was surprisingly agile.*

Below *Jack Brown, who usually works as an airline pilot, gets behind the stick of the P-47 for the Alconbury airshow.*

LOCKHEED P-38J LIGHTNING

Little is known of the operational history of The Fighter Collection's P-38, 42-67543. It was obtained by Marvin 'Lefty' Gardner in the early 1960s from an oil company in Oklahoma. The aircraft remained in storage with Gardner until 1988, when it was acquired by The Fighter Collection. In mid-1991, the aircraft was in final stages of complete restoration.

Unless the aircraft's original operational markings come to light in the interim, it is the intention to complete it as 'Happy Jack's Go Buggy', in the markings of Major Jack Ilfrey, USAAF, the first P-38 ace in the European theatre of operations.

Whilst The Fighter Collection has all the above mentioned aircraft either operating in flying condition on a regular basis or about to fly after restoration, other aircraft remain in the hangar at Duxford either undergoing or awaiting rebuild under the magic touch of the collection's dedicated engineers. These projects include the aforementioned ex-Thai Air Force

Right   *The only bomber in the collection, the* **B-25***, acts as a mother ship to the fighters during 'away' shows.*

Below   *John Romain, a driving force behind The Aircraft Restoration Company, takes time out to fly the* **B-25***.*

Bearcat and various Spitfires. In addition, a Russian designed Yak-11 once operated by the Egyptian Air Force with the serial 533 and an ex-RAF and Swedish Air Force North American Harvard IIB trainer, serial Fv.16105, will one day be rolled out of the hangar to take to the air once more.

The Collection also has a British-built Bristol Beaufighter XIC recovered from South East Asia, where it saw action against the Japanese. It was acquired in a complete but dismantled state and the long painstaking task of returning it to flying condition has commenced. When completed, it is likely to be the world's only flying example of this most versatile and respected adversary from the Second World War.

Such is the calibre of The Fighter Collection's engineering team that the Royal Air Force Museum chose them to restore to static condition an example of the Hawker Tempest II and this aircraft has recently been transported to its new home at Hendon.

DORNIER BÜCKER Bu 133 JUNGMEISTER

G-AYSJ, constructor's number 38, was one of a batch of Jungmeisters manufactured by Dornier for the Swiss Air Force immediately before the Second World War, and it remained with the Swiss as a trainer until the 1960s, serialled U-91. It then became HB-MIW on the Swiss civil register before coming to England as the mount of James Gilbert, founding Editor of *Pilot* magazine. He flew it in pseudo Luftwaffe training markings. The aircraft was exported to Germany in the early 1980s and came into The Fighter Collection in 1990.

It is now painted to represent the lead aircraft of the Luftwaffe's famous pre-war aerobatic team. Jungmeisters were the advanced fighter for the Luftwaffe and remained world class aerobatic competition aircraft until the 1950s.

Above right *'Grumpy', one of Walt Disney's seven dwarfs, adorns TFC's* Mitchell.

Right *The Bücker* Jungmeister.

Opposite *Quite a slender machine, the high summer sun reveals the B-25's construction and paintwork.*

THE OLD FLYING MACHINE COMPANY

For some years after the Second World War, the aviation industry in Great Britain still generated a great deal of excitement as many new types of aircraft emerged from the factories of the different companies that were eventually to be swallowed up by British Aerospace or to fall by the wayside. Even the average man in the street could name the test pilots of the day — men like Neville Duke, John Derry and Peter Twiss. Nowadays, most aviation enthusiasts would find it difficult to name the industry test pilots. To some extent, their place has been taken by the air display pilot. Arguably the best known are Ray Hanna and son Mark, who run The Old Flying Machine Company from Duxford.

Ray Hanna, a New Zealander, came to England to join the Royal Air Force after initial flying training in his native country in the late 1940s. Towards the end of a varied service career in which he flew many jets, including the Meteor, Vampire and Javelin as well as the piston-powered Hawker Tempest, he led the now world-famous Red Arrows from 1966, by which time they had become firmly established as the premier Royal Air Force aerobatic team.

Shortly before leaving the service, Ray began to display the Supermarine Spitfire Mk IX, then owned by Adrian Swire of Cathay Pacific Airlines. Then, after flying airliners for some years, including a period with Cathay, Ray, by now well known on the airshow scene for his superb displays of aerobatics and low flying, founded The Old Flying Machine Company to operate vintage aircraft for air display and film work.

Mark Hanna was taught to fly by his father at the age of sixteen, and joined the Royal Air Force two years later. During his time in the service, he flew the Jet Provost, Hawk, Hunter and Phantom. More recently, he left the RAF to concentrate on the activities of The Old Flying Machine Company.

The Old Flying Machine Company has been Duxford based for some years and, in that time, the collection of aircraft operated has expanded considerably. There has been some degree of change in the types flown with some such as the Stearman, Pilatus P-2, Harvard and P-40 Kittyhawk no longer being in the fleet. A brief period of operating World War One types also ended when these aircraft passed on to new owners,

Right *Classic foes. Pictured during the fiftieth anniversary of the Battle of Britain, the* Spitfire *and the* 109J, *once deadly rivals, fly together past the White Cliffs of Dover.*

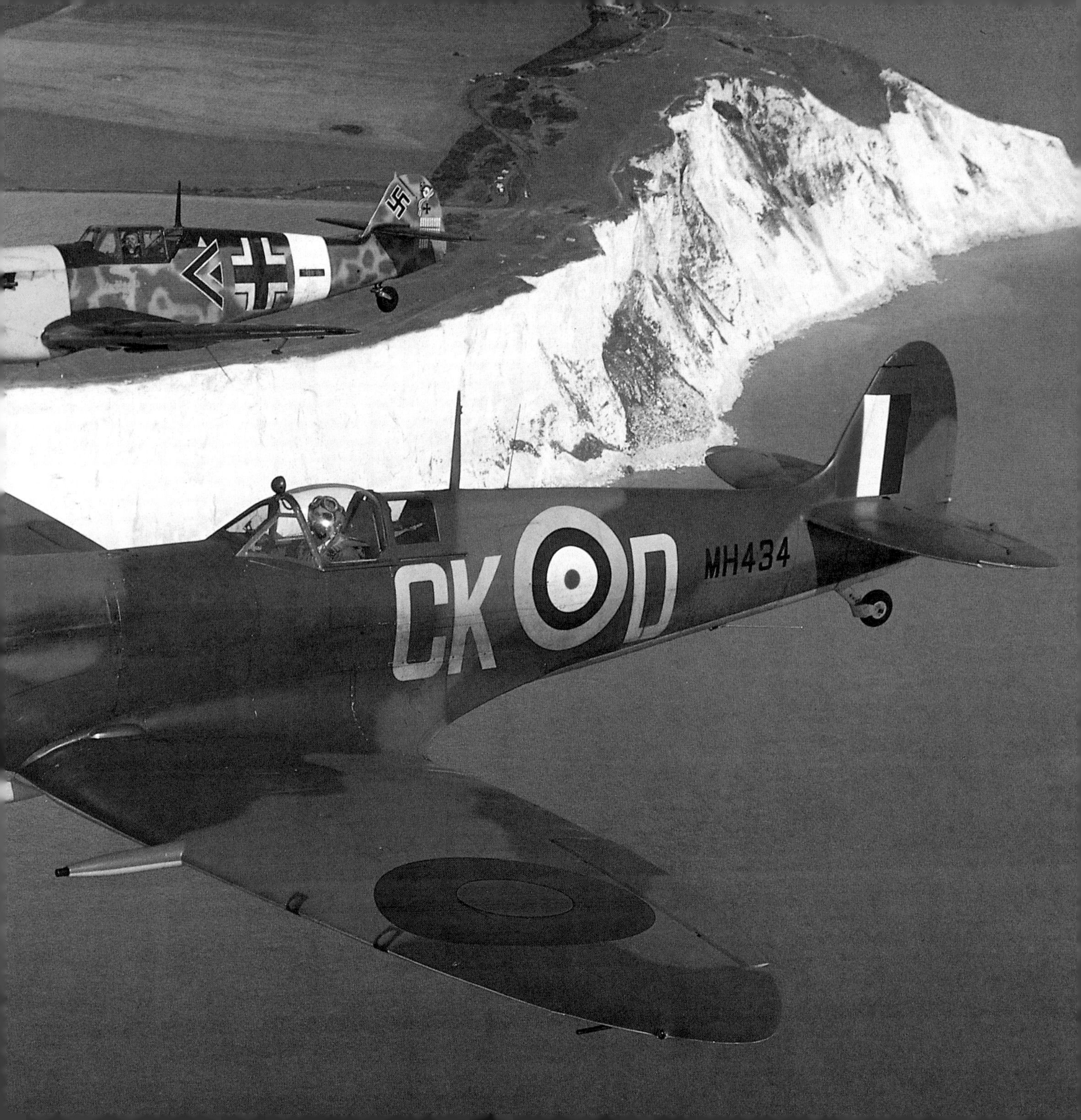
CK D
MH434

mainly based at North Weald. The early 1990s have seen the acquisition of several new types which are described later.

In addition to flying at air displays throughout Europe, The Old Flying Machine Company is also very active in the film industry, not only as flying participants — men and machines — but also as technical advisors. Films including *Empire of the Sun, Memphis Belle* and *Air America,* and TV productions such as *Perfect Hero* and *Piece of Cake* have all featured major contributions from The Old Flying Machine Company. They are also responsible, along with The Fighter Collection, for organising the Classic Fighter Display, a highlight of the Duxford display calendar.

At the end of 1991, The Old Flying Machine Company were operating the following aircraft . . .

SUPERMARINE SPITFIRE LF.IXB

The Old Flying Machine Company Spitfire is not only the collection's longest-serving aircraft, but is also one of the earliest privately owned

Right *R. J. Mitchell's beautiful elliptical wing design housed a formidable punch with a mix of 20mm cannon and .303 machine guns.*

'warbirds' to fly at British air displays. Indeed, it pre-dates the expression 'warbird' by some years! However, in the case of Spitfire G-ASJV, the name is not an idle boast as it saw active service during World War Two resulting in the destruction of several enemy aircraft.

Built originally in 1943 as MH434 at Castle Bromwich with the constructor's number CBAF.IX.552, it was to fly with three different RAF units, namely 222 (Natal), 350 and 349 Squadrons.

With 222 (Natal) Squadron, it was based at Hornchurch and flown by Flight Lieutenant H. P. Lardner Burke, DFC. Flying MH434, he shot down a Fw 190 on 27 August 1943, damaging another during the same sortie. On 8 September, another Fw 190 was claimed, together with a share in a Bf 109 three days later. MH434 was loaned to 350 Squadron for a few months in early 1944 before returning to 222. There followed a period out of the limelight with 84 Group Support Unit, during which time some accident damage was sustained. It then served for a while with 349 Squadron. In May 1945 it was placed in storage with 9 MU at Cosford, later moving to 76 MU at Wroughton.

In 1947 came a sale to the Royal Netherlands Air Force and service in the Dutch East Indies as H-105 with 322 Squadron, becoming H-68 in 1948. It saw active service in the counter-insurgency role against Nationalist forces in Indonesia. After a belly-landing caused by hydraulic failure, H-68 was shipped back to Holland, overhauled by Fokkers and test flown as serial B-13 before being sold again, this time to the Belgian Air Force as SM-41 in 1953. In 1956, it was disposed of to the Belgian organisation COGEA for use as a target tug, registration OO-ARA.

In 1963, Tim Davies purchased the Spitfire and it returned to England, becoming G-ASJV. 'Juliet Victor' was painted in a striking civilian colour scheme of blue, white and silver and was operated by Davies from Elstree until acquired by Spitfire Productions for use in the film *Battle of Britain*. After its period of stardom, it was purchased by Adrian Swire and flown regularly by Ray Hanna. It was painted in RAF camouflage with the serial MH434 and fuselage codes 'AC-S', Swire's initials.

In April 1983, G-ASJV was auctioned at Duxford and has since been operated by The Old Flying Machine Company. It has featured in several films and TV programmes and appeared in various different colour schemes according to role. These have included its original 222 (Natal) Squadron livery of RAF camouflage and 'invasion' stripes coded 'ZD-B', a PR grey scheme for a Hercule Poirot TV mystery, later being modified

by the addition of Norwegian Air Force colours, RAF camouflage and codes CK-D for *Perfect Hero* and, in the summer of 1991, Belgian Air Force colours.

MESSERSCHMITT 109 J

The Old Flying Machine Company is fortunate to be able to operate one of the Spitfire's adversaries in their collection also. In fact, G-BOML is a Hispano license-built version of the Messerschmitt Bf 109G-2 equipped with a Merlin engine. In Spanish service, the type was known as the C4K Buchon and the particular example flown by The Old Flying Machine Company was serialled C4K-107 and may have been constructor's number 170. Although not confirmed, it is believed that it flew with 7° Grupo de Caza-Bombardeo, possibly in the Sahara.

What is known for sure is that in 1967, Hamish Mahaddie saved many Buchons from oblivion by using them to represent Bf 109s in the *Battle of Britain* motion picture, C4K-107 amongst them. Although not actually used for any aerial sequences, it was used in a supporting role before

Right   *Notorious for its tricky ground handling characteristics, more 109s were lost to take-off and landing accidents at one stage than to enemy action. Brian Smith is pictured here taking off on his first solo on type.*

Below   *Ray and Mark Hanna taxi back after two consecutive demanding display routines.*

Overleaf   *Mean and magnificent. The 109 was one of the truly great fighters of the Second World War.*

Right and Below *Visibility from the 109 was generally poor. A serious disadvantage in combat, which it never really overcame.*

being exported to a museum in Illinois. After a period of storage, it was registered to Gordon Plaskett as N170BG and a restoration was commenced. Before completion, the project was acquired by the late Nick Grace, the aircraft arriving in England in 1986. It flew again on 6 May 1988 as G-BOML and was used in the TV film *Piece of Cake*.

The Buchon took up permanent residence at Duxford at the end of 1988 after Nick Grace's tragic death in a car accident. Since then it has appeared in several films and is a regular participant in set piece 'dogfights' at air displays.

CCF HARVARD IV/A6M2 ZEKE

The Hispano Buchon is not the only 'enemy' aircraft in The Old Flying Machine Company collection. During 1990 they acquired the converted Harvard N15798, formerly operated with The Harvard Formation Team at North Weald. This aircraft had started life in the early 1950s on the Fort William production line of The Canadian Car and Foundry Company where, as c/n CCF4-153, it was built for the Royal Canadian Air Force as 20362.

Right and Below *A recent acquisition, the OFMC's Bücker Jungmann was designed as a two-seat primary training machine. Mark Hanna gets back to basics in this highly aerobatic little biplane.*

Much later, it was substantially modified to resemble a Mitsubishi A6M2 Zeke or Zero for the famous *Tora! Tora! Tora!* film, one of several ex-RCAF Harvard IVs so modified. The structural alterations included the removal of the rear portion of the cockpit canopy and re-profiling of the fore and aft cockpit area, installation of a three-bladed propeller in place of the original two-blader, fitting a large propeller spinner, modified lower rudder, alterations to undercarriage legs and doors, a central bomb rack for dummy bombs, guns on the forward fuselage and a larger than original tail wheel.

When imported into the UK, N15798, as it had been registered for some years, was in a basic white colour scheme with red and white trim and pseudo Japanese markings, plus the tail code A1-110. After acquisition by The Old Flying Machine Company, the 'Zeke' was

Above *Ghost of the machine. Dave Southwood, Boscombe Down test pilot, slices through the overcast.*

Right Banzai! *The* Zeke *storms through a pyrotechnical battlefield during the North Weald Fighter Meet, 1991.*

repainted in a very authentic looking olive drab scheme with tail code 288 as operated from Rabaul in 1943.

COMMONWEALTH CAC-18 MUSTANG Mk 22

Amongst surviving Mustangs, The Old Flying Machine Company example is more unusual than most inasmuch as it is Australian built, having been produced under licence in 1951 by the Commonwealth Aircraft Corporation at Fisherman's Bend, near Melbourne. It left the production line as constructor's number 1517, Royal Australian Air Force serial A68-192, and first served with 1 Air Depot at Laverton, later moving to Tocumwal.

Upon disposal in 1958, the Mustang was privately owned as VH-FCB and based at Tamworth, subsequently moving to Moorabin with another owner. There followed a spell of air display work and air racing before eventually being shipped to a new owner in Manila, Philippines.

In 1969, it assumed the new identity PI-C651, but force-landed after

Right *Named after the wild prairie horse, the OFMC's* Mustang *aggressively closes in on the camera ship.*

Below *Flaring out on the approach to Runway 24 at Duxford, the* P-51*'s airspeed passes through 90 knots.*

Overleaf *Painted as 'Ding Hao!' (it means No. 1) for David Puttnam's motion picture* Memphis Belle, *Ray Hanna thunders skyward.*

engine failure on an early test flight. It did not fly again until August 1972, but in October 1983 it suffered engine problems again and considerable damage was caused upon landing back at Manila Airport.

By this time, the Mustang was officially registered as RP-C651. It was placed in store in a Manila warehouse, where it was purchased by Ray Hanna and shipped to Hong Kong in 1976.

Once in Hong Kong, a complete rebuild was carried out by HAECO (Hong Kong Aircraft Engineering Co), using some components from a former Philippine Air Force P-51, serial 44-72917/PAF410. In February 1985 the rebuilt aircraft arrived at Gatwick on board a Cathay Pacific Boeing 747. It became G-HAEC, having briefly appeared on the Hong Kong register as VR-HIV.

Once in England, it joined The Old Flying Machine Company and initially flew in RAAF markings with fuselage codes CV-H, representing an aircraft of 3 Squadron. When subsequently used in the filming of *Empire of the Sun* in Spain, it was repainted as *'Missy Wong of Hong Kong'*, tail code 592, later changed to 588. As such, it was modelled on a Mustang flown by the 118th Tactical Recce Squadron, 23rd FG, 14th Air Force of the USAAF in the Pacific at the end of the Second World War. It remained in those marks until its next film role in *Memphis Belle*. Then it was resprayed as 44-72917/AJ-A, *'Ding Hao'*, as flown by Major James Howard of the 354th Fighter Group. These markings have been retained through to the 1991 season.

GRUMMAN TBM-3E AVENGER

Grumman's portly torpedo bombing Avenger has been largely saved from extinction because of its post-war adoption by aerial fire-fighting and budworm-spraying outfits in the United States and Canada. By the time of their eventual withdrawal from these duties, there was sufficient awareness of the Avenger's historical value to ensure that many were purchased by 'warbird' collectors. The type has therefore become popular at air displays in North America and Europe with its surprising agility, speed and noise levels. The Old Flying Machine Company example is one of only two in Britain, and the only one fitted with a gun turret.

It was built at the Trenton, New Jersey, factory of General Motors' Eastern Aircraft Division in 1945, US Navy serial Bu.91110. Taken on strength in July of that year, it is known to have served with VA-22 in the

Above *Sunset sortie. Alan Walker runs up the Merlin before the long flight home.*
Right *Big friend, little friend, safely home.*

United States. Some time after being surplussed, it was operated in California as a 'sprayer' and, between 1963 and 1972, was flown by TBM Inc of Tulare, a well-known fire-fighting company. After passing through several more owners, it ended up with the Gro-Pro Corporation of Oklahoma from 1986 to 1988. During its civilian career, it operated with the registration N6827C, the identity it still bears now.

The Avenger arrived at Duxford in May 1988, and initially flew in Royal Navy markings with the spurious serial V110/BA. The Royal Navy operated many of the type both during and after the Second World War, calling them Tarpons until January 1944, when the type reverted to the name Avenger. In addition to air display flying, N6827C has also been used for air-to-air filming. In 1990 it was repainted to represent the US Navy Avenger flown during the Second World War by the future President of the United States, George Bush.

Right *The calm before the storm. The* TBM
*had an inauspicious start to its career when, on
its first mission during the Battle of Midway, five
of the six aeroplanes sent out from the* USS
Hornet *were shot down. The sixth returned, albeit
with a dead gunner, but had to be scrapped due to
battle damage. However, it went on to become
one of the most successful torpedo bombers of the
Second World War.*

Overleaf *Norwegian, Rolf Meum, on his first
solo in what he affectionately calls 'The Turkey'.*

CHANCE VOUGHT CORSAIR

Duxford has become something of a haven for this famous 'bent wing',
carrier-based fighter. In addition to the examples flown by The Fighter
Collection and Lindsey Walton, The Old Flying Machine Company had
two on strength by 1991.

The first, an F4U-4 model, was acquired in April 1988. It had
originally been built at the end of 1945 as c/n 9513 and flew with the US
Navy as Bu.97359, serving with VA-13 and VA-134 before embarking
on the aircraft carrier *USS Tarawa* for use in the Pacific around Japan and
Korea. By 1950, it had passed on to VMF-2, a US Marines unit, but
returned to the Navy and VA-44 some time after. Its active service life
ended in 1956 with entry into storage before purchase by Bob Bean of
Phoenix in 1959. Although registered as N5213V, it was not made
airworthy until 1976 when, as N97359, it flew for a US television series,

Above *'Big Hog', a Goodyear-built FG-1D, is the only ex-New Zealand Air Force Corsair still in flying condition.*

Right *The beauty of the beast. The mighty Corsair was the first US fighter to exceed 400 mph.*

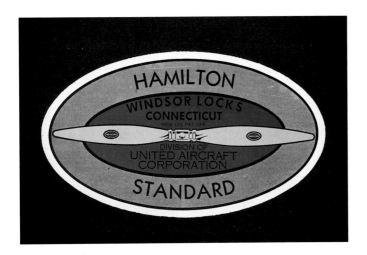

Baa Baa Black Sheep, based on the war history of Pappy Boyington. Re-registered as N240CA in 1980, it was owned for a time by Cinema Air of Houston and Merlin Air of King City, California.

N240CA initially flew with The Old Flying Machine Company in US Navy markings but was subsequently repainted as a Royal New Zealand Air Force machine with the serial NZ5628.

In 1991, The Old Flying Machine Company obtained a second Corsair, this time a Goodyear-built FG-1D model, c/n 3205. It had been built with the US Navy serial Bu.88391 but actually flew with the RNZAF as NZ5648, with whom it was on strength between August 1945 and May 1949. It escaped the scrappers and was displayed for a time at Hamilton Airport in New Zealand, then at a local garage, before arriving at Auckland's Museum of Transport and Technology. By 1972, it had found its way to Vancouver in British Columbia before passing through a succession of owners. During this time, it was registered as N55JP. In the late 1980s it was imported to the UK by Doug Arnold's Warbirds of

Right *Battle re-enactments are a major feature of many of today's air displays. Here, the* F4U-4 *scrambles to avoid the strafing runs unleashed by Gary Numan in his* Zero/Harvard.

Below *A lifelong problem with the* Corsair, *while landing on carriers, was the pilot's poor view over the massive nose. This situation was exacerbated when the cockpit on the production models was moved rearwards by three feet to make way for a 237 US gallon self-sealing fuel tank.*

Great Britain organisation. Named 'Big Hog', its colour scheme recreates that worn by the personal aircraft of Lieutenant Commander J. T. Blackburn, US Navy.

N55JP flew to a new lease of life with The Old Flying Machine Company at Duxford in early 1991. It will acquire RNZAF markings in due course.

YAKOVLEV YAK-11

One of the more unusual aircraft in The Old Flying Machine Company collection is the Yak-11, NX11SN. This two-seater aircraft arrived at Duxford from the United States in April 1990, and was soon re-assembled and test flown ready for that season's air shows. Notable for its short, stubby wings, this Russian designed aircraft was once flown by the Egyptian Air Force and was acquired from the former General Dynamics test pilot Neil Anderson.

Preceding pages   *Alan Walker, more usually found in the high tech, glass cockpit of an Air 2000 Boeing 757, is photographed here in the OFMC Yak 11, on a high level transit to RAF Finningley for their annual display.*

Right   *Shooting Star. Lockheed's magnificent* T-33 *jet trainer was actually faster than the* F-80 *fighter from which it was derived.*

Below   *Effortless grace. The T-Bird cuts through the cold, upper air, the domain of the new jet generation.*

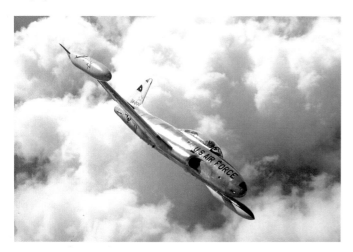

At the present time, the Yak is back in the land of its birth — Russia — being converted back to a single-seat Yak-3U.

HAWKER FURY FB.10

In mid-1991, an ex-Iraqi Air Force Hawker Fury FB.10 was obtained by the Hannas. One of a number of 'Baghdad Furys', it was imported into the United States in 1979 by well known American collectors Ed Jurist and David Tallichet. Its Iraqi military identity had been 243 but was replaced by the American civil registration N28SF. In 1982, it was shipped to Australia for Darwin-based Guido Zuccoli, who had it registered VH-HFX.

Once in Australia, the Fury was painted to represent a Nowra-based Royal Australian Navy Sea Fury WE729, fuselage code 115/NW. It later went through several Australian owners before coming to England.

CANADAIR T-33 SILVER STAR 3

Although The Old Flying Machine Company is noted for its collection of piston-engined aircraft, they also fly a jet — the Lockheed T-33, a classic design from the early days of jet-powered flight.

In the early 1950s, Canadair acquired a licence to build their version of the Lockheed T-33 with a Rolls-Royce Nene power plant. The Canadian designation was the Canadair CL-30 and it was named the Silver Star, although in common with users the world over, it became better known as the 'T-bird'. The Canadian forces used the Silver Star mainly for training purposes and examples were still being flown by them in the 1990s.

The Old Flying Machine Company T-33 was built as c/n 261, RCAF serial 21261, later 133261. Once retired, it was registered as CF-IHB to Ormond Haydon-Baillie and came to Duxford in late 1973, then becoming G-OAHB. It was painted in a spectacular black and white colour scheme, complete with a large knight's helmet on the forward fuselage and became well known in Europe for its flying displays prior to its owner's death in the crash of his Cavalier Mustang.

Subsequently sold, it became G-JETT, and after several changes of ownership both in this country and, subsequently, in American, it ended up in Switzerland as N33VC. It was from there that it was imported back into the UK in 1990 for a second residency at Duxford.

In addition to its airworthy collection of aircraft, The Old Flying Machine Company has several aircraft undergoing rebuild at the present

Right *Jet power!*

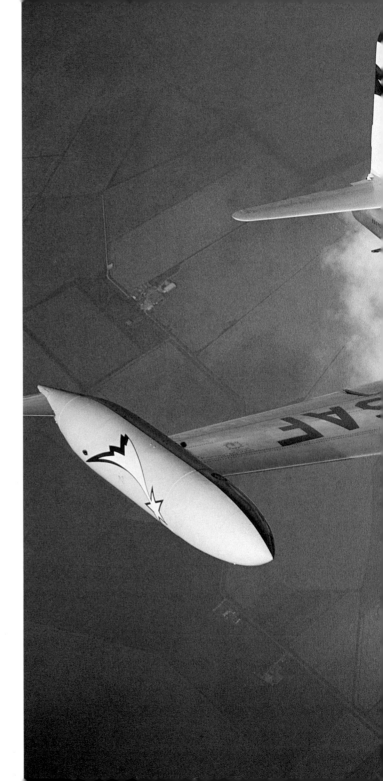

HOOF'S CUB

PIPER CUB

In America, they are called 'Warbugs'. They are the light liaison and training types used by the military and increasingly finding favour with the would-be warbird owner who cannot muster sufficient dollars to enter into the heavy trainer or fighter world.

In the UK, there are many Tiger Moths, an increasing number of Piper Club and Auster variants and a few Aeroncas, all in pseudo military markings. Nestled away amongst the heavy metal in the Fighter Collection hangar can usually be found a Piper Cub owned by Collection Pilot 'Hoof' Proudfoot, his two sons, and Martin Espin. Masquerading as a wartime Piper L-4 Grasshopper, it is painted up as a 'hack' aircraft used during World War Two by the Duxford-based 83rd Fighter Squadron, 78th Fighter Group who normally flew P-47 Thunderbolts. As can be seen from the photographs, the nose is adorned with sharksmouth markings as on the original aircraft.

This particular Piper was imported into the UK by well-known restorer Cliff Lovell and it was allocated the 'period' registration G-AKAZ in April 1982. These markings were unallocated back in 1947 when that sequence was in use! It had previously flown in France for many years as F-BFYL.

Right *Another successful father and son team in the UK, the Proudfoots. Here, Lee (Proudfoot, the younger) closes in.*

Below *The* Cub, *flown here by Hoof (Proudfoot, the elder) is painted in a genuine wartime Duxford scheme.*

CLASSIC WINGS

Above *Basic perhaps, but what a view.*
Right *Grace and style, as only de Havilland knew how.*

DE HAVILLAND DH 89A

Many visitors to Duxford may feel that, having seen so many magnificent old aeroplanes, they have a desire to fly in one. The Clacton Aero Club are able to satisfy such a desire in their magnificently restored de Havilland DH 89A Dragon Rapide. This classic biplane spends much of her time based at Duxford during the summer season and is used for pleasure flying.

This particular 'Rapide' was built in late 1943 and flew with the Royal Air Force as a wireless trainer. Known in RAF Service by the type name 'Dominie', she was serialled HG691.

Post-war, this lovely aeroplane was registered G-AIYR and passed through the hands of several owners, seeing use as a transport, photo-survey aircraft and as a parachutists' aerial platform.

In 1980, she was flown the 14,000 miles to Africa and back by David and Cherry Cyster to commemorate a similar pioneering flight by Sir Alan Cobham in 1925. Upon return, she was rebuilt by British Aerospace and was acquired by her present owners in 1991.

Lovingly cared for, G-AIYR is painted to resemble the similar aircraft of this type flown by Hillman's Airways from Clacton in the 1930s.

Another way in which the past can be sampled today at Duxford.

SPITFIRE Tr9
ML407/G-LFIX

A very recent addition to the ranks of flyable vintage aircraft at Duxford is Spitfire Tr.9 ML407/G-LFIX, made famous as the subject of the TV documentary, *The Perfect Lady*.

Although now a two-seater, ML407 was no cosy trainer to start with and it saw real action during World War Two. Built as an LF.IXC at Castle Bromwich in the spring of 1944, it was initially delivered to a Maintenance Unit at Lyneham prior to allocation to an operational squadron. Its first user was to be 485 (RNZAF) Squadron and it was delivered to their base at Selsey in Sussex in late May by ATA pilot Jackie Sorour, one of many female ferry pilots at that time. This is of interest, as will be seen later.

With 485 Squadron, MF407 flew as OU-V and it saw a total of 137 operational missions, mostly over occupied territory on the continent. During this period, its pilots claimed one Ju 88 shot down with a half share in a second, two Bf 109s destroyed and one damaged. At the very end of 1944, it briefly appeared on the strength of 341 (Free French) Squadron as NL-D before passing through the ownership of 308 (Polish) Squadron as ZF-P, 349 (Belgian) Squadron at GE-P, 485 Squadron again in February 1945, and 345 (Free French) Squadron as 2Y-A, complete with French roundels. Another nationality was added in April 1945, when ML407 became AH-B of 332 (Norwegian) Squadron.

With the war over, ML407 went back into storage but was sold back to its manufacturers in 1950 for conversion to a two-seater trainer. Briefly flown as G15-175, it became '162' with the Irish Air Corps, with whom it flew until 1960, then becoming an apprentice training airframe.

In the late 1960s, 162 came to England and was eventually acquired by Nick Grace for restoration at St Merryn in Cornwall. The rebuild involved retention of the two-seater format but with a redesigned rear canopy of more streamlined profile than the original. In fact, the result looked more like that used by the Russians on their small number of Spitfire trainers. Similar conversions have subsequently been used on two other UK-based Spitfires.

On 16 April 1985, ML407, now registered appropriately as G-LFIX, took to the air once more and was later painted in its original colours as OU-V of 485 Squadron. Since then, many former air and ground crews involved with this aircraft have been reunited with her. Tragically, Nick Grace was later killed in a car accident but ML407 continues to give pleasure to thousands, often in the hands of Nick's wife Carolyn, who was determined to fly it and in due course soloed in it, thus becoming ML407's second known lady pilot.

Left *The two-seat configuration was a big help in enabling Carolyn to become the only female pilot current on the* Spitfire. *She was trained by Pete Kynsey, who is pictured formating here prior to the Duxford Autumn Air Day, 1991.*

Right *Meticulously rebuilt by the late Nick Grace, this aircraft was the first Allied fighter to shoot down an enemy aircraft during the D-Day operations, in the days before its conversion into a two-seater.*

MESSERSCHMITT Bf 109G-2 'BLACK 6'

1991 saw the culmination of some nineteen years of hard graft by the restoration team involved in one of the longest running aircraft preservation projects in the UK. The aircraft which was the subject of all this care and attention is 'Black 6', a Messerschmitt Bf 109G-2, Wk.Nr.10639 and, at the present time, it is the only genuine Second World War combat aircraft of the German forces to be seen flying anywhere in the world. The 109's long rebuild is a story of dogged perseverance and eventual triumph which has resulted in a unique and truly exciting air show performer.

Black 6, so called because of its Luftwaffe fuselage codes, was built in Leipzig in late 1942 by Erla Maschinenwerk and went with III/JG77 to Cyrenaica and the Desert War. After only a few weeks, the Bf 109 was found combat damaged and abandoned at Gambut, Tobruk, by the Royal Australian Air Force. It was reflown by its captors, 3 Squadron, RAAF, and appeared adorned with their codes CV-V. Their plans to take it back to Australia were thwarted, however, when the RAF realised that this was a mark of 109 that they knew little about. After some comparative trials against various marks of Spitfire, Black 6 was shipped to England and joined 1426 Enemy Aircraft Flight at Collyweston. This unit was coincidentally a former Duxford inmate.

By early 1944, Black 6 had become RN228 in RAF camouflage and roundels and it continued to be used for evaluation and testing, both with 1426 and, later, with the Central Fighter Establishment at Tangmere.

Above *After an incredible nineteen-year rebuild, finishing at RAF Benson, the* Gustav *sinks gently onto the Duxford grass.*

Below *Currently the only Daimler Benz DB-605A engine flying in the world, ironically it was overhauled by Rolls-Royce. This is just one example of the enormous effort that Russ Snadden and his team have gone to, to preserve this award-winning thoroughbred's originality.*

Right *Return of the Eagle! Black 6, unleashed again into the skies over Britain. Flown here, in a strange twist of fortune, by Air Vice Marshal John Allison, RAF.*

Thereafter, it was placed in store, although was occasionally dusted off to appear at various open days. In 1961, Wattisham became home and an initial attempt at restoration to flyable status was started. Later restorers are agreed that these early efforts did little to improve the aircraft's condition.

Serious preservation commenced in 1972 when Russ Snadden, at that time serving with the RAF at Lyneham, took on the task of restoration with airworthiness in mind. Over the next two decades, Russ and his small but dedicated team worked against many odds in various locations — Lyneham, Northolt and, latterly, Benson — until in July 1990 the Daimler Benz DB605A engine was fired up for the first time, its rebuild having been completed by Rolls-Royce at Filton. Finally, in March 1991, Black 6 was flown again, this time from the grass at Benson and in the hands of Group Captain Reg Hallam.

The foregoing few lines cannot in any way do justice to the incredible project that led to that first flight in 46 years, but the accompanying photographs give a better idea of the quality of restoration.

The Bf 109G was known as the 'Gustav', and this fact is recognised by the registration G-USTV allocated to the aircraft. Some while before the

Above *Black death! Operated by The Imperial War Museum, the sleek and sinister shape of the 109 was often the last thing seen by unfortunate Allied aircrew.*

'first' flight, it was agreed that G-USTV would fly for a limited period and then go on show at the RAF Museum. Whilst flying, the aircraft would be operated by the Imperial War Museum at Duxford and in the latter part of 1991, the Messerschmitt was flown to her new base. Since then, she has dazzled crowds with her displays, often accompanied in mock dogfights by a Spitfire or two! She is usually flown by either John Allison or Dave Southwood.

Right *Recently rebuilt by Duxford neighbours, Historic Flying Ltd, David Tallichet's Spitfire XVI joins the 'Augsburg Eagle' in a unique display of Second World War might.*

THE RADIAL PAIR

NORTH AMERICAN HARVARD

The Radial Pair are one of the more recent arrivals to Duxford, basing themselves at the field in mid-1991. The team is one of very few specialising in warbird formation and synchronised aerobatics. Owners and pilots of the two Harvard aircraft making up the team are Gary Numan and Norman Lees, two of the most experienced Harvard pilots in Europe.

The Harvard is arguably the finest piston-engined trainer ever put into service. Difficult and demanding to fly well, over 15,000 were made in various marks for air forces all over the world. Powered by a 600 hp Pratt & Whitney 'Wasp' nine-cylinder radial engine, the Harvard can cruise at 150 mph and has a top speed in dives of 259 mph. With a maximum weight of 5,300 lb and a 42-foot wingspan, the Harvard is immensely strong, delightfully aerobatic and one of the most popular warbirds in the world today.

G-AZSC, a Harvard 2B, was built by Noorduyn Aviation, Montreal, Canada in November 1943. It was shipped to the UK for the Royal Air Force in January 1944 as FT 323 but then stored at RAF Lossiemouth until 1946, when it was sold to the Royal Netherlands Air Force. Retiring from

Right *How close can you get? Norman Lees and Gary Numan tuck into the camera ship.*

Below *Smoke go! Break go!*

military service in the mid 1960s, 'SC eventually returned to Britain and was subsequently bought by Gary Numan in 1984.

G-BDAM, another 2B, was also built by Noorduyn Aviation in 1943 for the Royal Air Force. 'AM served in Canada as part of The Empire Training Scheme until 1946, when it was transferred to the Swedish Air Force. In 1970 it was sold to its first civilian owner and entered the UK in 1973. It now wears the 1943 colours of No. 5 (Pilots) Advanced Flying Unit, RAF Tern Hill. The aeroplane is owned by Norman Lees and Euan English.

Left *Through the vertical. Guaranteed to get the adrenalin flowing.*

Right *Formation aerobatics is one of the most demanding disciplines in aviation. Here, Gary Numan leads The Radial Pair in a Line Astern Barrel Roll.*

THE AIRCRAFT RESTORATION COMPANY

Back in the early days of aircraft restoration at Duxford, one of the groups formed to rebuild and operate historic aircraft was The Fighter Display Team. Leading lights in this organisation were Graham Warner, Anthony Hutton and Robs Lamplough who, with an enthusiastic band of volunteers, gathered together a superb selection of aircraft, including Harvards, two Yak C-11s and a Spanish-built Messerschmitt Bf 109 or Hispano HA-1112 as it was rightly known. Later, two Beech 18 twin-engined transports were purchased and the organisation's name was changed to The British Aerial Museum of Flying Military Aircraft or British Aerial Museum as it became known.

In due course, Robs Lamplough moved on to North Weald and a new purpose-built hangar in which to house his growing collection. Likewise, Anthony Hutton also moved to North Weald and, on the opposite site of the airfield, created 'The Squadron' as a meeting point for vintage aircraft enthusiasts and operators and the home base for The Harvard Formation Team and, more recently, The Great War Combat Team. This left Graham

Right *Not in fact a* Fieseler Storch, *but a* Morane Saulnier Criquet, *a licence-built variant, flown here by John Romain.*

Below *Painted in winter camouflage typical of German aircraft operating on the Eastern Front, often in atrocious conditions, this rugged and reliable observation aeroplane stayed in service throughout the entire war.*

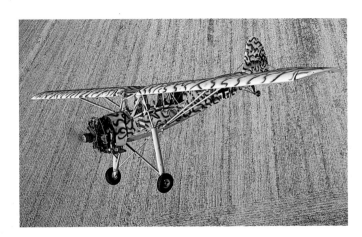

Warner and a small group of dedicated volunteers to look after his own aircraft fleet.

Today, The Aircraft Restoration Company, as the collection is now known, operates a variety of vintage aeroplanes but also carries out restoration and maintenance on machines owned by other organisations as opportunities arise and workflow permits. However, it is with one particular aircraft type — the Bristol Blenheim or 'The Forgotten Bomber'

as Graham Warner calls it — that The Aircraft Restoration Company has become synonymous. Their association with the Blenheim has been a long one, starting with high hopes and incredible ambition leading to triumph and then, amost immediately after, tragedy before coming full circle again with renewed determination to see a Blenheim airborne in British skies.

The Blenheim story really starts with ex-Royal Canadian Air Force Bolingbroke airframes that had originally been imported into this country by Ormond Haydon-Baillie in the mid-1970s. Bolingbroke was the name given to Canadian-built Blenheims. After Haydon-Baillie's death, his own collection of aircraft was dispersed and the Bolingbrokes were taken over by Graham Warner with the intention of continuing the restoration work that had commenced in order to produce one airworthy machine from available components.

The bulk of the project centred around Bolingbroke IVT serial 10038, originally built by Fairchild Aircraft Ltd of Loungueuil, Quebec, as one of a batch of 457 aircraft. Initially, work proceeded in one of the outbuildings alongside the main hangar complex and this workshop became known as 'Blenheim Palace'. 10038 had originally been discovered abandoned on a Manitoba farm and had been acquired by Haydon-Baillie from Wes Agnew. The rebuildings of this aircraft to flying condition took twelve years and in excess of 40,000 man hours to complete. The mainly volunteer team led by full-time engineer John Romain had to overcome the most extraordinary difficulties on the road to first flight, difficulties that were conquered by great patience and skill with, in some cases, industry help. This restoration project became a virtual legend in aviation circles and its completion became very eagerly awaited. Eventually, the end neared and the airframe was painted in RAF camouflage with the serial V6028 and the fuselage codes GB-D to represent a Blenheim IV of 105 Squadron. It was issued the highly appropriate civil registration G-MKIV, the only slightly less appropriate marks G-BLHM having not been taken up.

Triumph came on 22 May 1987, when V6028 made its 'maiden' flight

Left *1./JG 54's colours on the TARC* Fieseler Storch.

Right *Spirit of Canada. The* de Havilland Chipmunk *is arguably the finest post-war piston-engined trainer ever produced.*

from Duxford in the hands of John Larcombe and John Romain. However, cruel fate was only around the corner. A month later, on 21 June, the Blenheim left Duxford for an air display at Denham. During that show, the aircraft was lost after overshooting during a touch-and-go landing. In this much-publicised crash, the aircraft was destroyed, both engines having been ripped from the wings and the fuselage broken in two. Miraculously, all crew members escaped from the wreck.

Within a short period, Graham Warner and his team announced that they would somehow fly a Blenheim again. The Blenheim Appeal was launched and under the banner 'A Blenheim Will Fly Again', they have since worked toward this aim. A lesser team of individuals might have baulked at the challenge.

The 'new' Blenheim, boldly registered already as G-BPIV in readiness for its maiden flight, centres around another ex-Manitoba airframe, serial 10201, although some wing components from aircraft 9703 are being used. The shattered remains of V6028 have been stripped of useful components also. Bolingbroke 10201 arrived at Duxford in January 1988, having been purchased from aircraft collector Sir William Roberts' Strathallan Aircraft Collection in Scotland.

Work on this second project proceeds apace and amazing progress has already been made. John Romain, now a Co-Director of The Aircraft Restoration Company, still leads the project and is using the wealth of experience from rebuild number one to good effect. Since the early days, John has also become a qualified display pilot and flies many vintage aircraft at Duxford, not only for The Aircraft Restoration Company but for other organisations there also.

Although The Aircraft Restoration Company's Blenheim activity has tended to overshadow its other operations, their remaining aircraft are nonetheless of great interest and are frequently seen at airshows at Duxford and elsewhere. The other 'twin' in the collection is Beech 18 G-BKGL. The 'Twin Beech' is a sturdy aircraft originating in the mid-1930s and which enjoyed an incredibly long production run, the last example rolling off the line as recently as 1969! The type was widely used by the American forces but also saw worldwide military and commercial service. It was also known by the military as the Beech Expeditor. G-BKGL was built for wartime use but was rebuilt by the original manufacturers Beech in 1951 as a model -3TM. It served with the Royal Canadian Air Force from 1952, initially as serial 1564 and later as 5193, seeing a varied life all over Canada before being pensioned off in 1971. It was then placed on the Canadian civil aircraft register as C-FQPD and flew as a survey aircraft, latterly with Capital Air Survey Ltd. It was one of three Beech 18s sold from storage at Prestwick in Scotland and was acquired by the then British Aerial Museum, partly as a way of the team gaining some 'hands on' experience of operating a 'twin' in advance of completing the first Blenheim rebuild. For a while, it flew in US Navy colours of blue and grey as '164' until being resprayed during overhaul in 1989 as a silver US Army machine from 1942 complete with period 'meatball' roundels and striped rudders. It is often used as a photographic mount for air-to-air work.

All of the other aircraft operated by The Aircraft Restoration Company are single-engined. Possibly the most unusual one is the Morane Saulnier MS.505 Criquet G-BPHZ, more familiar to most as the Fieseler Storch. The Storch, an incredible looking machine with a long spindly undercarriage, high aspect ratio wings and an amazing low speed capability, was built in France as well as its native Germany during World War Two, Morane Saulnier building them under licence. G-BPHZ originated from this French production and flew with an Argus eight-cylinder inverted Vee engine and, as such, was known as a MS.500. After service with the French Armee de L'Air, it was re-engined in 1965 with a Jacobs radial engine, a modification

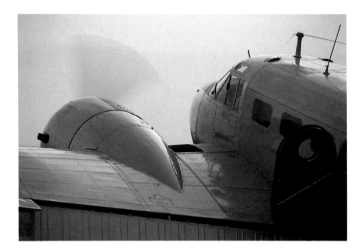

Above *Running up at dusk.*

Right *Looking immaculate in 1942 US Army Air Corps colours, the ARC* Beech Expeditor, *with its distinctive Wasp Junior radial engines, rumbles majestically through the summer sky.*

which did nothing for its looks or its noise abatement qualities! Re-designated as an MS.505 Criquet, it went on to serve as a glider tug registered as F-BJQC before lapsing into disuse in 1979. In 1988, it was acquired by The Aircraft Restoration Company, the somewhat less than pristine aircraft being flown to Duxford for rebuild. It emerged in splendid Luftwaffe colours representing a Storch of 1/JG54 as operated on the Eastern Front and with the 'fixed' fuselage codes of TA+RC reflecting its ownership.

The Aircraft Restoration Company have two 'yellow' aircraft, the first being the Auster AOP.9 G-AXRR/XR241. This aircraft was acquired at auction in 1983 having previously been operated by The Shuttleworth Trust at nearby Old Warden. Prior to that, it had been flown extensively by Major Mike Somerton-Rayner. In 1969, he had piloted this ex-Army Air Corps aircraft in the BP England to Australia Air Race, a challenge that was completed in sixteen days and over 140 flying hours. At present XR241 is painted in the yellow scheme it wore during that race rather than the camouflage in which it spent the majority of its military career and it resides at the Army Air Corps Museum at Middle Wallop on loan. The other yellow aircraft is DHC-1 Chipmunk 22 G-BNZC, which represents a Royal Canadian Air Force machine, serial 13671. This attractive aircraft, restored by The Aircraft Restoration Company immediately after the loss of the first Blenheim, originally flew with the RAF as WP905.

During the last couple of years, The Aircraft Restoration Company has operated the Fairey Flycatcher replica G-BEYB/S1287 on behalf of owner/builder John Fairey. This 'home-built' is based on the original design by John's father Sir Richard Fairey and, with the exception of the engine, is a faithful replica of the 'real thing'. As there were no Armstrong Siddeley Jaguar engines available, a more modern Pratt & Whitney had to be used instead.

Rebuild projects under way at present include Westland Lysander III G-LIZY for the Imperial War Museum. Originally it was intended that this would be a restoration to airworthy condition, but priority work on the Blenheim project has led to a change of plan. It is an ex-Canadian aircraft and has previously carried the identities Y1351, 1558 and V9300. Inspection of the airframe at Duxford reveals it to be yet another example of the team's superb rebuildings skills. Also 'on the go' is North American T-6G 41-32473. This is a complete rebuild utilising parts of other T-6s/Harvards, including the ex-Sandhurst dump KF487 and the RCAF painted G-BGPB which unfortunately came to grief in an accident at Gransden some while back. It is possible that a second T-6 may emerge from this project.

Amongst all this activity, The Aircraft Restoration Company also finds time to work on aircraft owned by other organisations. These have included the Imperial War Museum's P-51D Mustang 44-73979 now hanging from the roof at Lambeth as 44-72258 'Big Beautiful Doll' and two of their German aircraft, Focke-Wulf Fw 190A-8 Wk.Nr.733682 and Heinkel He 162A-1 Wk.Nr.120235. An aircraft from a different era was Hawker Hunter F.4 WN904, formerly in the Imperial War Museum Collection at Duxford but restored for exhibition as a 'gate guard' at RAF Waterbeach.

The team have also resprayed aircraft for The Old Flying Machine Company and worked on other aircraft from Stephen Grey's and Lindsey Walton's collections.

Below *TARC's second* Blenheim *under progression.*

B-17 PRESERVATION
L I M I T E D

Arguably the best-known privately owned 'warbird' in the world is Boeing B-17G Flying Fortress 'Sally B'. She has operated from Duxford since 1975 and is very much a part of the scene there.

'Sally B' was one of many Flying Fortresses to roll from the Lockheed Vega production line in Burbank, California. Built in 1944, she bore the constructor's number 8693 and designation B-17G-105-VE. She saw service with the USAF in the United States as 44-85784 and was later used for trials of infra-red equipment by the General Electric Flight Test Centre at Schenectady, New York. Modifications for this work included a re-profiled nose, whilst the installation on the starboard wing of a man-carrying pod may possibly have been in connection with other trials to test the feasibility of piloting the aircraft from such a position.

In the late 1950s, the 'Fort' was disposed of to the French organisation IGN — Institut Geographique National — at Creil, near Paris. Registered as F-BGSR, she was used extensively for photographic survey work around the world, one of several B-17s in IGN's fleet.

In 1975, IGN started to dispose of some of its B-17s and F-BGSR was sold to Ted White, who re-registered her as N17TE and had her flown to England. She took up residence at Duxford in March of that year and has remained there ever since. Initially, she was painted to represent an aircraft of the 749th Bomber Squadron, 457th Bomber Wing. The following year, the markings G-BEDF were allocated. Since then, 'Sally B', as she has become known, has worn several different colour schemes and appeared in various films and TV programmes, most recently the re-work of the famous movie *Memphis Belle*. A common factor in the recent liveries has been the black and yellow chequers worn on the starboard inner engine cowling. These are in memory of Ted White, who perished in a flying accident in Malta in June 1982. The Harvard in which he crashed had similar markings on its nose.

'Sally B' must be almost unique in the extent to which she is loved by her many fans. There can be few aircraft that have engendered so much public support. In addition to her popularity as a Duxford resident, she has been the central part of the annual Great Warbirds Air Displays at West Malling in Kent over the last ten years.

Right   *Classic nose art. During the summer of 1990, five* B-17s *operated once again from a Cambridgeshire airfield. The filming of the motion picture* Memphis Belle *starred Duxford's most famous resident,* B-17 *Preservation's* Sally B, *painted on this occasion as the* 'Belle' *herself.*

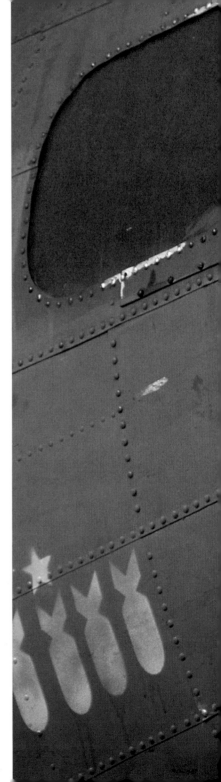

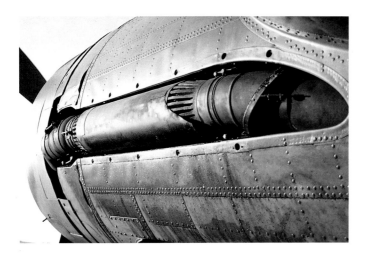

Above 'Sally B' *at rest.*

Right *With her chin turret removed to resemble an F model, 'Sally B' thunders overhead during the Classic Fighter Display, Duxford, 1991.*

Elly Sallingboe and B-17 Preservation Limited are the operators of 'Sally B' and she flies throughout Europe as a memorial to the USAAF in Europe and to Ted White, the man responsible for saving her for posterity. She is maintained and flown by a dedicated and experienced crew and supported by a large and enthusiastic society — The Sally B Supporters Club.

Right *A Labour of Love. 'Sally B' serves as a flying tribute to the 79,000 US airmen who lost their lives during the Second World War.*

Below *In 1944 the English countryside often echoed to the sound of a lone* B-17 *limping home. Its crew, tired and battle weary, willing their battered machine on. The* Flying Fortress *rarely let them down.*

LINDSEY WALTON COLLECTION

Lindsey Walton's 'day job' is farming potatoes on his land at Sutton Bridge in Lincolnshire. In his 'spare' time, he has become an accomplished air display pilot, well known and popular on the European air show circuit.

At the present time, he owns three very different vintage display aircraft, one of which is based at Duxford, the other two being regular visitors there.

Biggest of the three is also one of the longest serving high performance fighters to be seen at European air shows, having been imported to this country in 1982. This is his Chance Vought F4U-7 Corsair, NX1337A. For some years, it was the only Corsair flying in Europe and it has travelled far and wide in the hands of Lindsey and other display pilots, including John Allison and the late John Watts.

Although not built until around 1952, it is nonetheless a genuine warbird, having seen active military service. It was originally built for the United States Navy as serial Bu.133722 in Dallas, Texas, the latest in a line of Corsair variants that had previously distinguished themselves during World War Two and in Korea. The F4U-7, powered by the Wright R-2800-43W driving a four-bladed propeller, was built under the Mutual Defense Aid Programme for the French Navy or Aeronavale for use in Indo-China. In all, ninety-four of this version were built and delivered to the French. After use in Indo-China, 133722 went on to serve in Algeria and in the Suez Campaign in 1956. It was flown at various times by 15 Flotille as 15.F.3 and 14 Flotille as 14.F.12. By the early 1960s, the Aeronavale Corsairs had been withdrawn and Lindsey's aircraft returned to the United States after a period of storage at Toulon. Eventually, it was rebuilt by Gary Harris and painted in its attractive Aeronavale colours of dark blue and yellow 'Suez' stripes representing aircraft 15.F.22 of 15 Flotille.

Lindsey Walton purchased NX1337A, as she had become, in 1982 and she was flown across the Atlantic via Iceland, arriving in time to appear at some of that year's air shows. With the advent of other Corsairs in the United Kingdom, including some at Duxford, warbird fans have been privileged to witness Lindsey's machine in formation with others of the breed, a sight considered highly improbable not that many years ago.

Right *Twilight tails.*
Below *Lindsey tucks up the gear and powers the* F4U-7 *into the final Great Warbirds Air Display at West Malling.*
Overleaf *A true warbird. Lindsey's* Corsair *saw combat with the French Aeronavale in the Suez.*

LEA AVIATION & PLANE SAILING AIR DISPLAYS

Lea Aviation was formed some years ago for the purpose of operating vintage aircraft and staging the now famous Fighter Meet Airshow held annually at North Weald in Essex, only a few miles down the M11 motorway from Duxford. Later, the associated organisation Plane Sailing Air Displays came into being. Both operators were formed by three individuals well known in aviation circles — Royal Air Force pilots John Watts and Paul Warren Wilson and photographer Arthur Gibson. The rationale behind the Fighter Meet was to provide excitement and spectacle in a show organised by pilots who really knew about display flying whilst their aircraft acquisitions reflected a desire to fly unusual and distinctive types. Since 1987, both organisations have been based at Duxford, where they moved from Barkston Heath in Lincolnshire. Sadly, John Watts was killed in a tragic mid-air collision between two RAF Tornados in August of 1988, but his memory is perpetuated in the Fighter Meet and by 'his' aircraft displayed by close friend Paul Warren Wilson.

Right *Consolidated's PBY was a true flying boat, complete with hull and mast (to support the massive flapless wing).*

Below *Resplendent in its new paint scheme, representative of 210 Squadron, RAF, Paul Warren Wilson takes 'Killer Cat' low out of Duxford.*

CONSOLIDATED PBY-5A CATALINA

One of the larger vintage aircraft on the air display scene is Plane Sailing's Consolidated PBY-5A Catalina amphibian. The Catalina was built in greater quantity than any other flying-boat type and served in considerable numbers with not only the United States and British forces during the Second World War but also with the air arms of several Commonwealth

Right *Love it or loathe it, the* Catalina *is unarguably one of the most distinctiv shapes on the warbird scene.*

countries and those of the Soviet Union. Post-war, the Catalina fle worldwide with military and commercial operators, mainly in it amphibious version, and was used for forest fire fighting and aerial surve as well as freighting and passenger carrying.

Plane Sailing's aircraft was built as a PBY-5A amphibian in late 1944 b Consolidated in New Orleans and saw service, primarily as a trainin aircraft, with the US Navy until retirement in 1952. Its US Navy serial wa Bu.46633, whilst its production line number was 1997. After bein surplussed, it spent some time in the great US Navy graveyard at Litchfiel Park, Arizona, before being sold in the civil market in late 1956. With th ferry registration N10023, it left Arizona and was subsequently sold to Canadian company who wished to use it as an executive transport. I identity was changed to CF-MIR.

The new owners decided to follow the example of some other post-wa PBY operators and convert the Catalina to a 'Super Cat' with 1700 h Wright Cyclone R-2600-35s replacing the original 1200 hp Pratt & Whitneys. In addition, the nose turret and fuselage 'blister' turrets wer removed along with most of the wartime interior equipment whilst a larg angular rudder replaced the original curved unit. The work was carried ou in Canada by Noorduyns and the finished conversion was known as 'Super Canso S/C.1000'. There followed many years of operation i Canada, the aircraft eventually being used for photographic and geo physical survey work. For a while, it was operated with the America identity N608FF before returning to the Canadian register in 1970. Its la Canadian owner, Geoterrex, used the Catalina in South Africa until it 'retirement' in 1984.

Left *Gear down, checks complete. The* Catalina *established on long finals, in this case for the tarmac of Duxford.*

Plane Sailing Air Displays acquired the Catalina and flew it back to the UK in February 1985, and after being repainted in RAF markings it embarked on a new lease of life at air shows throughout Europe. Although registered as G-BLSC, the markings worn are those of 210 Squadron's JV928/Y, the aircraft in which Flight Lieutenant John Cruickshank earned his Victoria Cross.

Based at Duxford since mid-1987, the 'Killer Cat', as it is affec-tionately known, is now a familiar sight to many and has its own enthusiastic support in the form of The Catalina Society. The 1990 display season saw its former glory partially restored with the installation of superbly rebuilt blisters on the rear fuselage. When possible, the opportunity is not missed to operate from water, and over the last few years 'Killer Cat' has got its bottom wet at several locations, not only in England but in Sweden, Norway and Northern Ireland.

Right   *Barely wider than the pilot it held, the* Tigercat's *rearward visibility in combat left a lot to be desired.*

Below   *This sleek thoroughbred is powered by two Pratt & Whitney R2800s, giving it over 4,200 hp!*

GRUMMAN F7F-3 TIGERCAT

The second 'twin' in the collection is a real 'hot ship' and very much a contrast to the rather stately Catalina! One of the most powerful aircraft to come out of the Second World War and too late to take part in it, the Grumman Tigercat is a sleek thoroughbred and the only example of its type still flying outside of the United States. Indeed, it is most likely that Lea Aviation's F7F is the only one ever to have flown the Atlantic, the two that flew here just after the war for evaluation having been shipped across.

Grumman F7F-3 Tigercat, Bu.80483, was built in mid-1945 for the United States Marine Corps but saw very little service before being struck off and consigned to storage at Litchfield Park. It was saved from the scrapman and was used as a high performance forest fire fighter in California for a number of years, flying with Cat-Nat Airways and Sis-Q Flying Services with the registration N6178C. It lapsed into disuse in the mid-1980s.

John Watts had set his sights on acquiring a really 'different' warbird, and the quest was satisfied when 78C was discovered at Santa Rosa in California. Sadly, John was not to live to see his dream realised, but Paul Warren Wilson eventually acquired the aircraft and delivered it by air to Duxford in an epic ferry flight that traversed the whole of the United States and then the Northern Atlantic.

Upon arrival at Duxford, the Tigercat was repainted in US Navy colours complete with tail codes 'JW' in memory of John Watts. Exciting as a solo

Left *Used as a night fighter in Korea, the* **Tigercat** *is seen here transiting home from RAF Lakenheath.*

Right *The only example of the* **F7F** *flying outside of the United States, it was flown across the Atlantic to the UK by Paul Warren Wilson in November 1989.*

performer, Paul has also flown her in company with other 'warbird' types including a quartet of Grummans at the 1991 Classic Fighter Display along with The Fighter Collection's Bearcat and Hellcat and Tony Haig-Thomas' TBM Avenger.

PILATUS P-2

The third aircraft in this small Duxford-based collection is the Pilatus P-2 piston-engined trainer G-BJAX. Originally built by the Swiss company Pilatus to their own design but incorporating some components from the Messerschmitt Bf 109, all those built served exclusively with the Swiss Air Force. When the type was retired, the Swiss held an auction and aircraft U-108, a P-2.05 variant, was purchased for The Old Flying Machine Company. It initially flew in markings representing a Swiss operated Bf 109 with the serial J-108. Later acquired by Lea Aviation, it was repainted to resemble a Luftwaffe fighter of the famous Richthofen Geschwader JG2.

The Pilatus has seen use in many mock battles between Allied and enemy types at shows and has also proved its worth as a solo performer and photographic mount.

Above and Right   *The late John Watts, co-founder of The Fighter Meet display and Plane Sailing, captured here flying with the spirit and skill that made him one of Britain's most popular performers.*

Below   *Mike Carter taxies the* P2 *in after returning from Leicester's annual air display. Mike, ex Red Arrow and Harrier instructor, was the P2's display pilot for 1990.*

BAC WINDOWS
YAK - 11

YAKOVLEV YAK-11

The Yak -11 is the last offshoot of a long line of impressive Yakovlev fighters of the Second World War, the wing and basic construction of the metal and fabric fuselage being quite similar.

Designed as an advanced combat trainer, it was the first aircraft to be used by the expanding satellite forces of the Soviet Union in the 1950s and some 4,557 were built, 707 of those being produced in Czechoslovakia.

First flown in 1946 and entering service only a year later, the Yak-11 is powered by a 820 hp Shvetsov ASh-21 seven-cylinder radial engine, and has a maximum speed of 370 mph.

For a while, in the 1960s, it was a transitional aircraft between the basic trainers of the time and the two-seat MiG 15UTI jet.

G-OYAK is an ex-Egyptian Air Force, Czech-built C-11, sponsored by BAC Windows and owned by their chairman, Eddie Coventry. It flew again in 1991 after a four-year rebuild by Phil Parrish and his team, and currently wears the markings of Ivan Kozhedub's Lavochkin La 7, the top scoring Soviet ace of World War Two with sixty-two victories.

Above   *Blue sky, white wings. Poetry in motion, Soviet style.*

Left   *Gary Numan tucks in.*

Right   *Russian thunder! One of the most potent piston-engined trainers ever built, the* Yak-11 *flown here by Norman Lees.*

Overleaf   *Ghost: a warbird lives and breathes high over Duxford.*